Springer Series on

ATOMIC, OPTICAL, AND PLASMA PHYSICS 55

Springer Series on

ATOMIC, OPTICAL, AND PLASMA PHYSICS

The Springer Series on Atomic, Optical, and Plasma Physics covers in a comprehensive manner theory and experiment in the entire field of atoms and molecules and their interaction with electromagnetic radiation. Books in the series provide a rich source of new ideas and techniques with wide applications in fields such as chemistry, materials science, astrophysics, surface science, plasma technology, advanced optics, aeronomy, and engineering. Laser physics is a particular connecting theme that has provided much of the continuing impetus for new developments in the field. The purpose of the series is to cover the gap between standard undergraduate textbooks and the research literature with emphasis on the fundamental ideas, methods, techniques, and results in the field.

Please view available titles in *Springer Series on Atomic, Optical, and Plasma Physics* on series homepage http://www.springer.com/series/411

See Leang Chin

Femtosecond Laser Filamentation

With 58 Figures

Springer

See Leang Chin
Center for Optics, Photonics & Laser
and
Dept. of Physics, Eng. Phys. & Optics
Université Laval
Pavillon d'Optique-Photonique
QUÉBEC QC G1V 0A6
CANADA

ISSN 1615-5653
ISBN 978-1-4939-5076-8 ISBN 978-1-4419-0688-5 (eBook)
DOI 10.1007/978-1-4419-0688-5
Springer New York Dordrecht Heidelberg London

Softcover re-print of the Hardcover 1st edition 2010

Printed on acid-free paper

Springer is part of Springer Science+Business Media (www.springer.com)

To my wife, May.

Preface

This book attempts to give a discussion of the physics and current and potential applications of the self-focusing of an intense femtosecond laser pulse in a transparent medium. Although self-focusing is an old subject of nonlinear optics, the consequence of self-focusing of intense femtosecond laser pulses is totally new and unexpected. Thus, new phenomena are observed, such as long range filamentation, intensity clamping, white light laser pulse, self-spatial filtering, self-group phase locking, self-pulse compression, clean nonlinear fluorescence, and so on. Long range propagation at high intensity, which is seemingly against the law of diffraction, is probably one of the most exciting consequences of this new subfield of nonlinear optics. Because the intensity inside the filament core is high, new ways of doing nonlinear optics inside the filament become possible. We call this filamentation nonlinear optics.

We shall describe the generation of pulses at other wavelengths in the visible and ultraviolet (UV) starting from the near infrared pump pulse at 800 nm through four-wave-mixing and third harmonic generation, all in gases. Remotely sensing fluorescence from the fragments of chemical and biological agents in all forms, gaseous, aerosol or solid, inside the filaments in air is demonstrated in the laboratory. The results will be shown in the last part of the book. Through analyzing the fluorescence of gas molecules inside the filament, an unexpected physical process pertaining to the interaction of synchrotron radiation with molecules is observed. It is the excitation of the superexcited states of a molecule which undergo dissociation into neutral fluorescing parts. Thus, there is a similarity between exciting the superexcited states of a molecule through multiphoton/tunnel excitation and through one XUV-photon absorption. This phenomenon will be described in the last chapter.

Also included in the last chapter is the most recent discovery and development of filament induced anisotropy in air. Normally isotropic air becomes birefringent because of the strong field of the linearly polarized pump pulse. A probe pulse will experience the ultrafast birefringence through cross-phase modulation. Molecular rotational wave packets of air molecules are excited and aligned; they then dephase and revive after the pump pulse is long gone. The ease in detecting such revivals using a probe pulse is indeed a surprise.

Some background knowledge of the readers is assumed. This includes basic nonlinear optics, the properties of femtosecond laser pulses, and multiphoton/tunnel ionization/excitation of atoms and molecules.

Quebec, Canada S.L. Chin
July 9, 2009

Acknowledgement

This book is the result of many years of collaborative research work with my students and postdoctoral researchers as well as with many national and international colleagues from many countries. It would be very difficult to mention all of them without running the risk of forgetting some. I thus choose not to list their names. I do hope that they understand me. I would like to thank Prof. Gordon Drake who encouraged me to write this book and took a lot of time to read the manuscript giving precious suggestions for improvement and making many language corrections.

Contents

Chapter 1
Introduction

This book discusses the consequence of the so-called self-focusing of a very short laser pulse in a transparent medium. Self-focusing is similar to focusing of a light/laser pulse by a lens except that this lens is a self-created one. If a laser pulse has an intensity distribution across the wave front and if the intensity is strong at the center and weak at the edge (a Guassian distribution in intensity, for example), the high intensity central part of the wave front would see a higher index of refraction in the medium due to nonlinear increase of the index of refraction than the low intensity edge. This is similar to focusing a plane wave by a lens where the central part passes through a thicker glass medium than the edge. Because the velocity of propagation is equal to c/n where c is the velocity of light in vacuum and n is the index of refraction of the lens' material, the light wave thus propagates through the central part of the lens slower than through the edge resulting in the curvature of the wave front toward the axis of propagation. This means focusing. The same is true in self-focusing except that this is a continuous self-inflicted process throughout the propagation. It will thus come to a self-focus where ionization occurs. Different sections/slices of the pulse along the propagation direction self-focus at different positions along the propagation axis giving rise to a continuous series of self-foci which we call filament. This description will be expanded in more detail in Chapter 2.

When using/propagating femtosecond (10^{-15} s) laser pulses in any transparent optical medium, there is no way to avoid self-focusing of the pulse so long as it is sufficiently intense. When it is powerful enough, self-focusing will collapse resulting in plasma generation. In the olden days, using long laser pulses, this phenomenon was and still is a nuisance one likes to avoid inside laser systems. But with femtosecond (fs)laser pulses, the phenomenon becomes what is now popularly called filamentation with a lot of potential applications.

1.1 Mature Physics and New Development

This is a story of the revival of a so-called mature field of science. Nonlinear optics was developed following the discovery and development of lasers in the 1960s.

S.L. Chin, *Femtosecond Laser Filamentation*, Springer Series on Atomic, Optical, and Plasma Physics 55, DOI 10.1007/978-1-4419-0688-5_1,

Phenomena such as self-focusing and self-trapping of laser beams in an optical medium (Hercher, 1964; Chiao et al., 1964), multiphoton/tunnel ionization (see for example Chin and Lambropoulos, 1984 and a recent review by Chin, 2004) and optical breakdown (see for example: Morgan, 1975; Raizer, 1991 and references therein) of materials by a laser pulse at high intensity were well known. Filamentation and spectral (red) broadening were observed by Reintjes et al. (1973), while the limitation of the laser intensity at the self-focus (which we now call "intensity clamping") in glass was proposed already in the 1970s (Yablonovitch and Bloembergen, 1972). Probably no one would have predicted that a powerful femtosecond laser pulse could uncover a seemingly known yet totally unexpected sub-field of applied physics, namely, filamentation nonlinear optics (Chin et al., 2007). Special names, some of which became popular, were given to describe known or unexpected phenomena. Examples are filamentation, intensity clamping, background (energy) reservoir, self-transformed white light laser pulse, self-spatial filtering, self-group phase locking, self-pulse compression (down to the few cycle and even single cycle levels), clean nonlinear fluorescence, and so on. Long range propagation at high intensity which is seemingly against the law of diffraction, is probably one of the most exciting consequences of this new sub-field of nonlinear optics. Short range propagation in glass resulting in the transformation of its properties is also as fascinating. The main reason for the observation of the above mentioned new phenomena is the very short duration (femtosecond) and high power/intensity of the laser pulse. Nanosecond or even picosecond laser pulses could not have produced such results because of optical breakdown that masks the observation of the new phenomena.

Imagine the following scenario. A femtosecond laser pulse propagates and self-focuses into a small volume in air at a high intensity such that some molecules are ionized. It then keeps on propagating in this form as if it were a bullet through a long distance up to the kilometer range. A streak of plasma column is left behind by the bullet; the streaking "bullet" gives rise to the perception of a filament. As it propagates, its frequency keeps on broadening "by itself". Soon, it turns white (very broadband) encompassing what is popularly called supercontinuum. (See for example: Alfano, 1989 and references therein.) It is called a chirped white light laser pulse. This self-transformed white light laser pulse could be considered as a white light bulb streaking in the sky whose back illumination (scattering) could be applied to molecular absorption spectroscopy, hence identifying some molecules in air at long range.

Inside the core of this filament, the high intensity can explode chemical and biological molecules/agents. Many of the neutral fragments emit fluorescence with a characteristic spectrum; i.e., each species will emit its own finger print fluorescence spectra. Because we can in principle project the filament at long distances in air, we can thus detect chemical and biological agents in air from a long distance using the laser radar (LIDAR) technique. Solid targets interacting with the strong filament can be identified at long range through what is called femtosecond laser induced breakdown spectroscopy using this remote sensing technique.

Associated with this fundamental bullet of high intensity laser light is another light bullet at the third harmonic frequency of the fundamental bullet, born through the interaction of the fundamental bullet with air. This third harmonic bullet is dragged along by the fundamental bullet at the same speed of propagation (self-group phase locking). The third harmonic UV pulse as well as the high intensity fundamental pulse could excite fluorescence from chemical and biological molecules at long range. Sodium in air could be excited to fluoresce, hence forming an artificial "star". The filament could in principle conduct electricity, hence providing lightning control. High spatial quality few cycle (down to single cycle) intense pulses tunable from the UV to the radiofrequency could in principle be generated inside the filament in gases by mixing it with another pulse at an appropriate frequency.

Imagine again the propagation of such pulses in glass. The filament is short. Inside the filament, the glass material is melted by the very short femtosecond pulse in an initial interaction time much shorter than the heating time. When the melted material re-solidifies, a new zone of different index of refraction is born. This leads to the so-called writing of a waveguide or generating data storage space in the material.

Many of these are previously unheard-of phenomena. The author believes that more new phenomena, deeper physical understanding and new applications are waiting to be discovered. He attempts to describe and explain the basic physics and applications of this new field of filamentation nonlinear optical science principally in air (gases), while that in condensed matters will be briefly described. The book aims at those who have an interest to enter field. The presentation of this book is at the level of graduate students in physics and researchers from other fields.

During filamentation of a femtosecond laser pulse, some fundamental nonlinear processes are involved in the propagation and interaction, namely nonlinear phase changes of the laser pulse, multiphoton and tunnel ionization, and optical breakdown. These concepts will be explained briefly in the rest of this chapter. The most popular femtosecond Ti-sapphire laser (central wavelength around 800 nm) will be used in our discussion as the laser source throughout this book.

1.2 Phase Effect of a Laser Pulse Propagating in an Optical Medium

A laser pulse is an electromagnetic pulse. Mathematically, it can be represented as a Fourier integral by the superposition of a set of plane waves propagating in the z-direction in the following form, for example

$$E(z,t) = \int_0^\infty E_0(\omega,t)\exp\left[i\left(\omega t - kz\right)\right]d\omega \qquad k = \omega n/c \tag{1.1}$$

The plane wave component in the integrand contains the wave vector k which is expressed as a function of the index of refraction n. It is this index that plays a central role in the propagation, self-focusing and filamentation of a femtosecond laser pulse. This is because n is dependent on many parameters. In the linear regime, it is wavelength dependent. In a transparent material, under normal dispersion condition, longer wavelengths have a lower index. Thus, a plane wave at a longer wavelength (say red color) will travel faster than that at a shorter wavelength (blue). The initial coherent superposition of many waves at different wavelengths gives rise to a short pulse. However, once they start propagating, the red will propagate faster than the blue components and the pulse becomes longer.

Many good books deal with the aspect of the superposition of wave and ultra short pulses. The readers are referred to the following, namely, Rullière C ed. (1998) and Diels and Rudolph (2006).

1.3 Multiphoton and Tunnel Ionization

Multiphoton ionization (Delone and Krainov, 1994; Chin and Lambropoulos, 1984) is simply an extension of the phenomenon of single photon ionization. This is a bound-free transition. In the context of the perturbation theory, an electron in a bound state in an atom or a molecule, after absorbing one (usually) energetic-photon, would have the probability of being excited into the free (electron) state or continuum. In the case of condensed matter, it is the ejection of an electron by the energetic-photon that overcomes the work function. In both these situations, the photon energy is larger than the ionization potential or the work function. This is the normal photoelectric effect in which one photon is absorbed.

What if the photon energy is less or much less than the ionization potential or the work function? Quantum mechanics, in the framework of the perturbation theory (Goeppert-Mayer, 1931), shows that if the intensity of the radiation is high enough, there will be a distinct probability that the atom or molecule or the condensed matter absorbs several photons "simultaneously" (in one cycle of the field oscillation) and excites a bound-free transition thus releasing a free electron; this is probable when the total energy of the absorbed photons is larger than the ionization potential or the work function. In Fig. 1.1a, a schematic picture of multiphoton ionization of an atom or a molecule is shown. The absorption of several photons simultaneously overcomes the ionization potential, giving rise to the probability of ejecting an electron. This picture can also represent the case of releasing a free electron from the surface of a metal (condensed matter) where now the work function is overcome resulting in the ejection of a free electron. In the case of filamentation in a condensed matter which is transparent to the incoming femtoseocnd laser pulse, a multiphoton transition of an electron from the valence band to the conduction band is more probable (Brodeur and Chin, 1998, 1999). This is shown in Fig. 1.1b. The electron set free into the conduction band will be further accelerated by the laser field through collisions resulting in a breakdown, but only partially in the case of femtosecond laser pulses (see below).

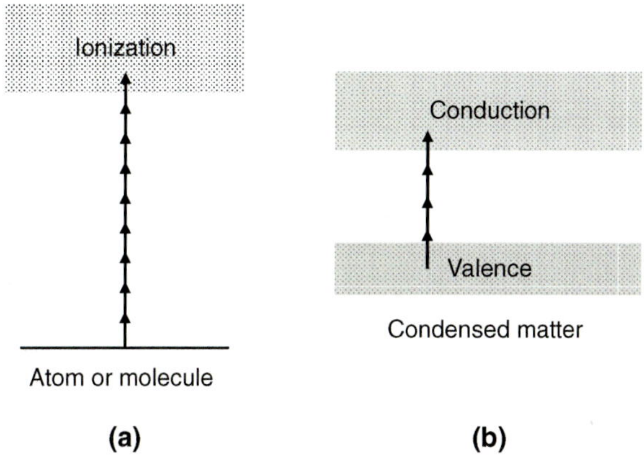

Fig. 1.1 Schematic diagram illustrating multiphoton transitions. (**a**) Multiphoton ionization of an atom or a molecule; (**b**) multiphoton transition from the valence to the conduction band in the case of a transparent condensed matter

However, when using femtosecond laser pulses to ionize gases, it is more appropriate to talk about tunnel ionization (Fig. 1.2). Tunnel ionization of gas atoms or molecules is similar to multiphoton ionization in the sense that both are high intensity effects. Depending on the combined condition of the laser and the ionization

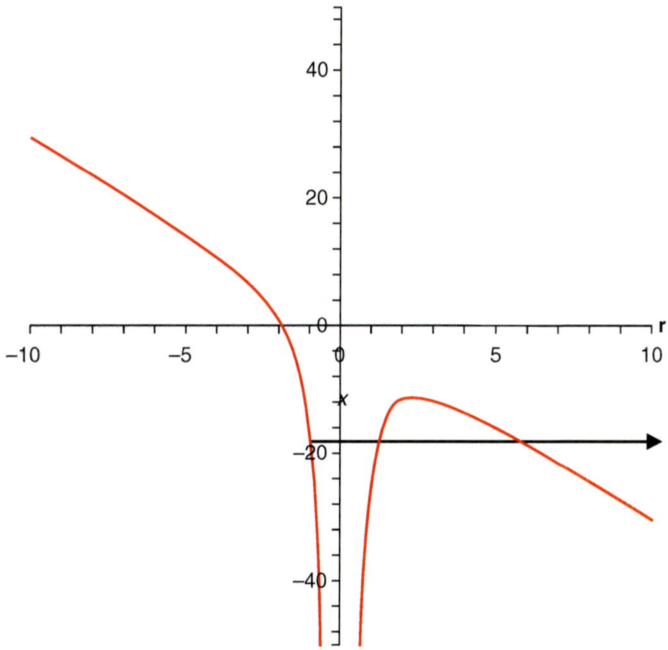

Fig. 1.2 Tunnel ionization

potential of the gas particle to be ionized, we could be in the multiphoton or the tunnelling regime. One popular parameter that distinguishes between multiphoton and tunnel ionization is the Keldysh (adiabatic) parameter γ (Keldysh, 1965). The first observation of tunnel ionization of atoms and molecules was by Chin (1984) and Chin et al. (1985), after more than 20 years of debate by those in the field. From our experimental findings, when $\gamma < 1/2$, it is in the tunnelling regime and when $\gamma < 1/2$, it is in the multiphoton regime (Chin, 2004). For a more detailed discussion of multiphoton and tunnel ionization, see Chin (2004).

1.4 Optical Breakdown

Since the invention of the Q-switched laser at the beginning of the 1960s, optical breakdown in air or condensed matter became an easily observable physical phenomenon. When focusing any intense nanosecond or picosecond laser pulse in the material, optical breakdown is bound to occur. Before filamentation was well understood, optical breakdown was a rule with no known exception.

It was started by Meyerand and Haught (1963, 1964) who observed that a focused Q-switched ruby laser pulse could generate a spark (plasma) in air. After a great deal of extensive theoretical and experimental work in many laboratories, it was finally proposed that optical breakdown of gases followed essentially a three-step process (Fig. 1.3) as follows: (1) Multiphoton ionization (MPI) of impurity molecules with low ionization potentials would easily provide a few free electrons with low initial kinetic energy in the focal volume at the front part of the pulse. (2) The free electrons in the strong laser field could absorb or emit n photons ($n = 0, 1, 2, 3\ldots$)

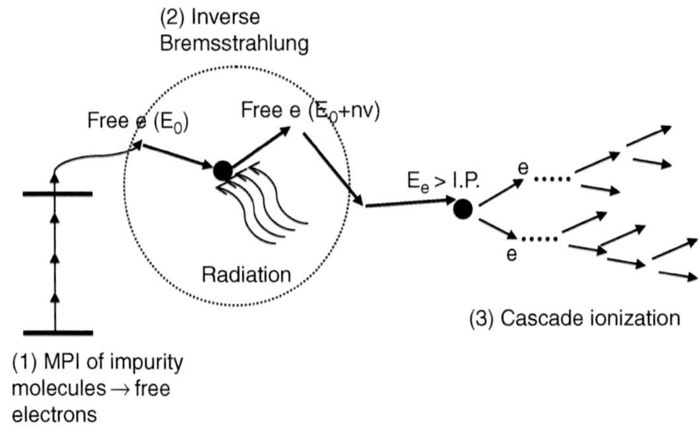

Fig. 1.3 Visualization of laser induced breakdown, a three steps process: (1) MPI of impurity molecules, (2) inverse Bremsstrahlung resulting in electron acceleration, and (3) cascade ionization

while colliding (scattering) with a much heavier particle (atom, molecule or ion). The heavy particle is to conserve momentum during the interaction. The process of absorbing n photons is called inverse Bremsstrahlung or free-free transition. (3) After one or more inverse Bremsstrahlung processes, the free electron would acquire a kinetic energy E_e higher than the ionization potential of the gas molecule/atom. Subsequent collision would give rise to the ejection of an extra electron from the molecule/atom. This would result in two low energy electrons. They would undergo the same processes as before each giving rise to two more electrons, and so on, until the gas is fully ionized. This is called cascade or avalanche ionization; i.e., breakdown (Morgan, 1975). Many experiments were performed to verify the cascade ionization process (Morgan, 1975). The fundamental process of the initiation of breakdown by the MPI of impurities was proved experimentally by Chin (1970) whereas the free-free transition which is at the center of cascade ionization was observed by Weingartshofer et al. (1977, 1979).

In condensed media transparent to the laser pulse, the interaction leading to breakdown inside the material is essentially similar to that in gases except that the first few "free" electrons are generated through multiphoton absorption by these electrons from the valence to the conduction bands. Once in the conduction band, these electrons would again undergo inverse Bremsstrahlung followed by cascade ionization resulting in the breakdown of the material (Raizer, 1991).

One major difference between breakdowns in gases and in condensed matter is that the density of atoms/molecules in a condensed medium ($\sim 10^{22}$ cm^{-3}) is about three orders of magnitude higher than that in a gas, say, at one atmosphere ($\sim 3 \times 10^{19}$ cm^{-3}). Thus it is much easier to induce breakdown in a condensed medium than in gases. In fact, using femtosecond laser pulses, practically no inverse Bremsstrahlung and cascade (avalanche) ionization could take place in gases at one atmospheric pressure at the intensity of $\sim 10^{13}$ W/cm^2. This is because the mean free time of a free electron released from an atom or a molecule through multiphoton/tunnel ionization is longer than the pulse duration. The mean free time of a free electron is given by the formula (Raizer, 1991)

$$(\Delta t)_{\mathrm{mean}} = \frac{1}{\sigma N_a v_e} \tag{1.2}$$

where $(\Delta t)_{\mathrm{mean}}$ is the mean free time; σ, the electron-neutral collision cross section; N_a, the atomic/molecular density; v_e, the free electron velocity in the field.

We estimate the mean free time as follows. The clamped intensity in air is $\sim 5 \times 10^{13}$ W/cm^2 (Kasparian et al., 2000a; Becker et al., 2001). According to Paulus (2007, private communication), from the observed electron spectra of the above threshold ionization (ATI) (Becker et al., 2002; Milosevic et al., 2006) of atoms at the intensity range of 10^{13}–10^{14} W/cm^2, the probability of an electron to become really free is more than 99%. That is to say, not more than 1% of the ionized electrons participate in the rescattering process (Corkum, 1993; Kuchiev, 1987, 1995, 1996). Thus, rescattering of the electrons could be neglected. The free electron will experience an oscillatory motion in the laser field acquiring a mean kinetic energy (known as the ponderomotive potential) given by (units are given in the parenthesis) (Augst et al., 1991)

$$\frac{e^2E^2}{4m\omega^2}(\mathrm{eV}) = (9.33 \times 10^{-14})I(\mathrm{W/cm^2})\lambda^2(\mu\mathrm{m}) \quad (1.3)$$

where e and m are the electronic charge and mass respectively; ω, the laser frequency; E, the electric field strength of the laser pulse; I, the laser intensity and λ, the wavelength. It is between 0.6 and 6 eV in the intensity range of 10^{13}–10^{14} W/cm^2. Its nonrelativistic velocity is thus $v_e \sim 4.5 \times 10^7$ to 1.4×10^8 cm/s. N_a of one atmospheric air is 2.68×10^{19} cm^{-3}; $\sigma \sim 10^{-15}$ cm^2 for both gas and condensed matter for electrons with kinetic energy ranging from 1 to 6 eV (Raizer, 1991). We obtain $(\Delta t)_{\mathrm{mean}} \sim (3\text{–}8) \times 10^{-13}$ s or 300–800 fs. This is longer than most of the femtosecond laser pulses used in the experiments.

In the case of condensed matter, the density is 1000 times higher than that of air at one atmosphere. Thus, the mean free time is 0.3–0.8 fs. Sun et al. (2005), using 130 fs Ti-sapphire laser (800 nm) pulses, have measured that the electron collision time (mean free time) inside the filaments in fused silica is of the order of 1.7 fs. The theoretical estimation is in good qualitative agreement with the experimental result. The order of magnitude of mean free time in condensed matter could thus be roughly 1 fs. This would mean that if a whole 50 fs (at FWHM) pulse focuses into a small focal zone, there would be a few tens of cycles of collisional ionization resulting in optical breakdown. This is true when focusing the laser pulse by a very short focal length lens. However, because of slice-by-slice self-focusing (see Chapter 2), the number of collisions is much less when using a long focal length lens to focus the fs pulse into a condensed medium. We shall come back to this question in Chapter 2.

1.5 Intense Femtosecond Laser Beam Attenuation

Very often, one needs to vary the energy of the powerful femtosecond laser pulse. The classical technique is to use transmission-type attenuators. There would be some modification of the physical properties of the pulse (wave front, pulse duration, chirp, etc.) after passing through the attenuator because of nonlinear effects (self-focusing, GVD, etc.) (Tang et al., 2005a,b). To avoid this, one could use reflective attenuators. Another way is by adjusting (rotating) a half-wave-plate in front of a polarizer. The half-wave-plate is oriented such that it will rotate the polarization of the uncompressed chirped laser pulse which the fixed polarizer will transmit the selected polarization, thus attenuating the laser pulse. The half-wave-plate/polarizer pair is placed before the compressor. If the uncompressed pulse is very energetic (10 mJ or higher), it is wise to place them in the path of the seed beam before the latter is further amplified. This latter technique of using the half-wave-plate/polarizer pair ensures that the beam energy is adjusted before the compression of the pulse; it thus avoids the deterioration of the wave front of the pulse through absorption optics due to nonlinear propagation effects. This latter technique is adopted in all the author's experiments in Laval University. However, for experiments involving a window of a vacuum system or using lenses, the same problem would arise. In the work of Tang et al. (2005a,b), this problem was discussed thoroughly. A good way is to use the half-wave-plate/polarizer pair in front of the amplifier as mentioned above and put the compressor and the beam steering optical elements and the interaction zone all inside the *same* vacuum. In the case of propagation experiment in optical media, the compressed pulse has to get out of the vacuum system and the window will always be a problem. It is a question of the tolerance of the particular experiment under study.

Chapter 2
Filamentation Physics

2.1 Some Experimental Observations

The basic physics of filamentation is universal and occurs in all transparent media (gases, liquids and solids). After propagating through an optical medium, a femtosecond (fs) Ti-sapphire laser pulse (at around 800 nm) turns into a white light laser pulse whose transverse pattern shows a central white spot surrounded by colored rings. The only difference is that the length of the filament is different in different media while the free electron generation mechanisms inside the filament core are different between gases and condensed matter materials. Figure 2.1a–d show the evolution of the transverse patterns of a 5 mJ/45 fs/800 nm Ti-sapphire laser pulse after propagating and filamenting in air without external focusing. The transform limited pulse from the vacuum compressor propagates into a 10 m vacuum pipeline which is connected directly to the vacuum compressor (see Fig. 2.2). After passing through the 1 cm-thick CaF_2 exit window, the pulse enters the corridor next to the author's laboratory. It then passes through an inverted telescope so as to reduce the beam diameter from about 3.5 cm to about 6 mm. The diameter is measured at the $1/e^2$ level of the fluence distribution of the beam pattern on a white paper taken by a CCD camera. The colors in the single-shot-pictures in Fig. 2.1 are real. They are patterns of the pulse intercepted on a piece of white paper and are taken by a digital camera. The distance z in the picture indicates the distance to the white paper screen from the exit of the inverted telescope (Fig. 2.2). This is a manifestation of single filamentation of the pulse that self-transforms into a white light laser pulse. The colored rings are conical emissions.

When the energy of the pulse becomes higher, multiple filaments occur. Figure 2.3a–f shows a series of similar patterns of the Ti-sapphire laser pulse at 50 mJ/45 fs propagating in air without the inverted telescope. The initial beam diameter at $1/e^2$ level of the fluence distribution is around 3.5 cm. The distance z indicated in the figure is the distance measured from the beam steering optics (high reflectivity mirrors, see Fig. 2.2) which is set just outside the exit window of the pipeline. Because of the shot-to-shot fluctuation, the single shot patterns do not reproduce themselves, but the general trend of change is obvious. Each hot spot tends to self-focus into a white light pattern but there is also a filament competition process

S.L. Chin, *Femtosecond Laser Filamentation*, Springer Series on Atomic, Optical, and Plasma Physics 55, DOI 10.1007/978-1-4419-0688-5_2,

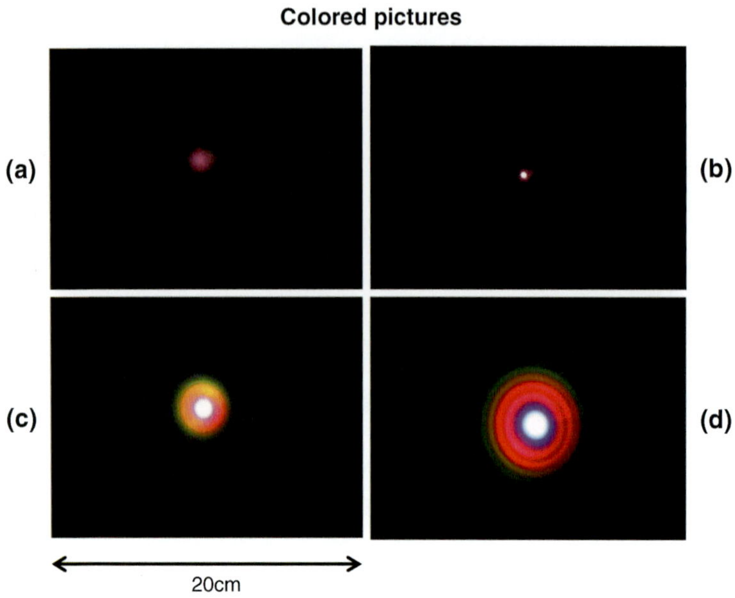

Fig. 2.1 Single filamentation. These are pictures taken by a normal digital camera

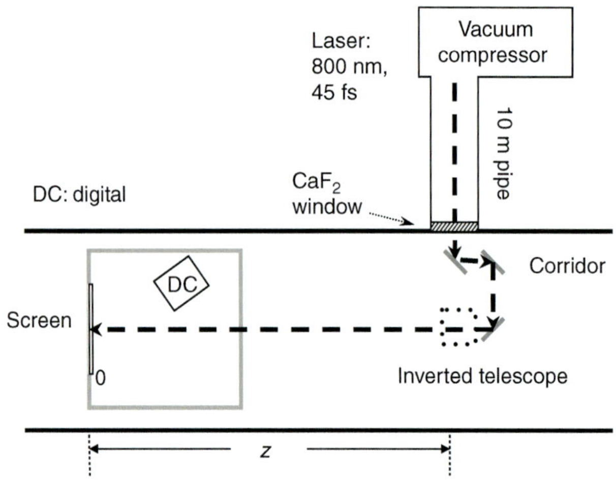

Fig. 2.2 Experimental setup to measure the evolution of the patterns of laser pulse during filamentation in air over long distances

among the hot spots to become white. This will be discussed later in the book under filament competition. The colored star like lines in Fig. 2.3e and f indicate the interference between fields of conical emissions originated from different self-focal spots (filaments). We shall later discuss how such a star like object can be created.

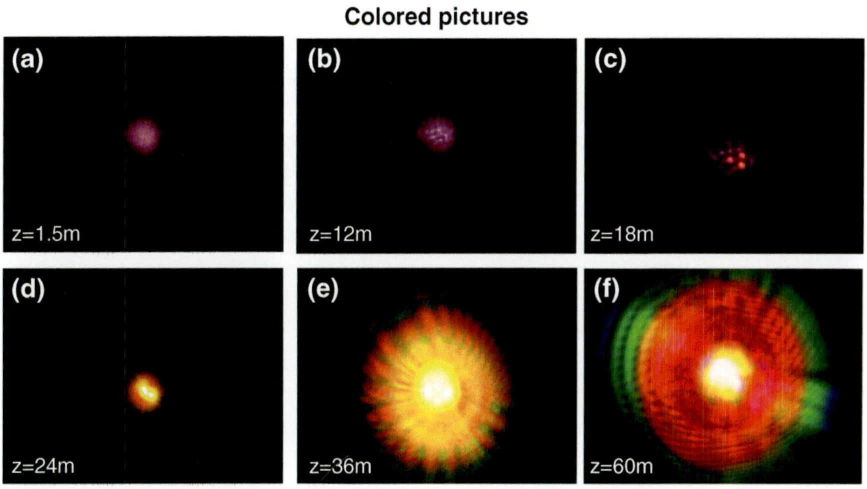

Fig. 2.3 Multiple filamentation. The horizontal scale of each picture frame is 20 cm. These are colored pictures taken by a normal digital camera

Such colored patterns can also be easily observed from the propagation in a condensed medium using a much lower laser power. Figure 2.4 shows an extreme case of propagating an unfocused Ti-sapphire laser pulse at 8 mJ/40 fs (diameter $\sim$ 6 mm at FWHM of the fluence distribution) through a piece of 4 mm thick BK7 window. The peak power of the light is about 0.2 TW. This is about 5 orders of magnitude higher than the critical power for self-focusing in glass (2–3 MW). It thus generates

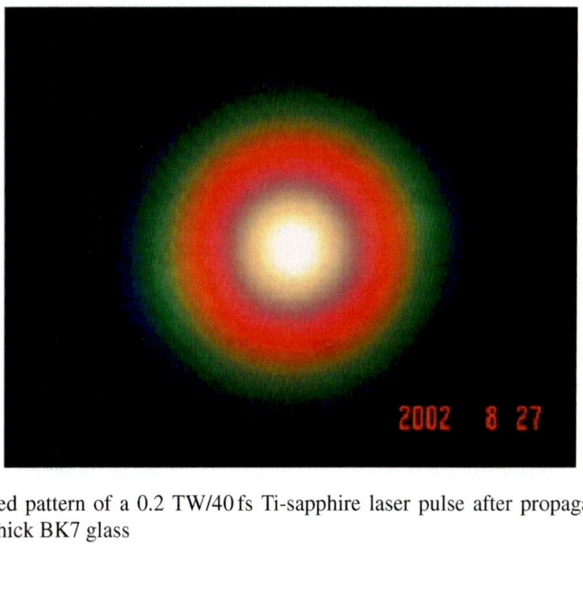

Fig. 2.4 Colored pattern of a 0.2 TW/40 fs Ti-sapphire laser pulse after propagating through a piece of 4 mm thick BK7 glass

a large number of closely spaced filaments each becoming a white light laser source. Each source evolves into a pattern similar to that shown in Fig. 2.1c, d. The ensemble of these tightly spaced white light sources thus shows a more diffused set of colored rings (conical emission) with no resolution.

Now, when we look at the filament from the side in air, we can barely see a faint white line. Figure 2.5 shows a picture taken with an ICCD (intensified CCD) camera from the side of the filament which is the result of the propagation of a 45 fs Ti-sapphire laser pulse in air in a clean laboratory environment (class 100,000). The fine line of light comes mainly from the fluorescence of nitrogen molecules (Talebpour et al., 2000, 2001; Becker et al., 2001a). The diameter of the line is less than 100 μm. This is a manifestation of a single filament; but the line is not uniform. It shows a series of brighter sections separated by darker zones. This is a manifestation that the laser pulse undergoes multiple refocusing, generating a few filaments along the same propagation axis.

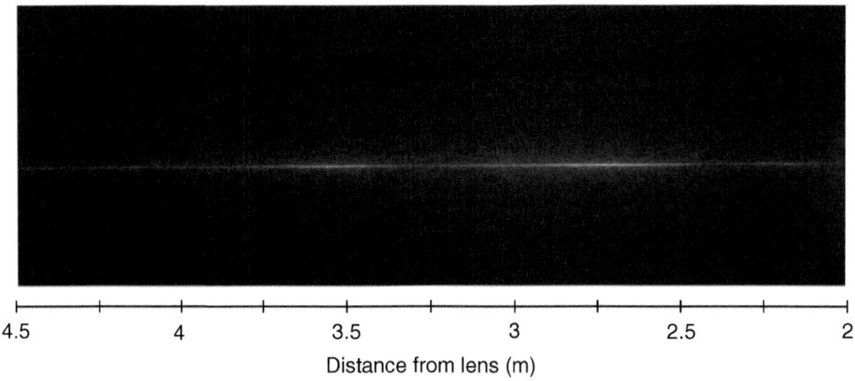

Fig. 2.5 Side view of a filament in air. The diameter is 70 μm. It shows re-focusing

If we use a sensitive burn paper to intercept the pulse at different positions of the propagation, we will observe a pattern similar to the one shown in Fig. 2.6. The central black spot is the so-called self-focal spot and the succession of such spots (self-foci) along the propagation axis gives rise to the perception of a filament. Because inside this hot spot, the intensity is high (about 5×10^{13} W/cm^2; see Kasparian et al., 2000a), nitrogen molecules are tunnel ionized. Tunnel ionization and multiphoton ionization are highly nonlinear processes that can be observed only at high laser intensities (see for example: Chin, 2004 and references therein). Thus, the low intensity pedestal outside the hot spot would not give rise to any measurable ionization signal. After the laser pulse has passed, the ionized molecules in the series of hot spots (self-foci) relax through collisions and fluoresce, giving rise to the picture shown in Fig. 2.5. It looks as if there was only a single filament (a series of hot spots) along the propagation axis; but in fact, there is a lot more radiation energy stored in the surrounding area which is normally not observed or omitted in

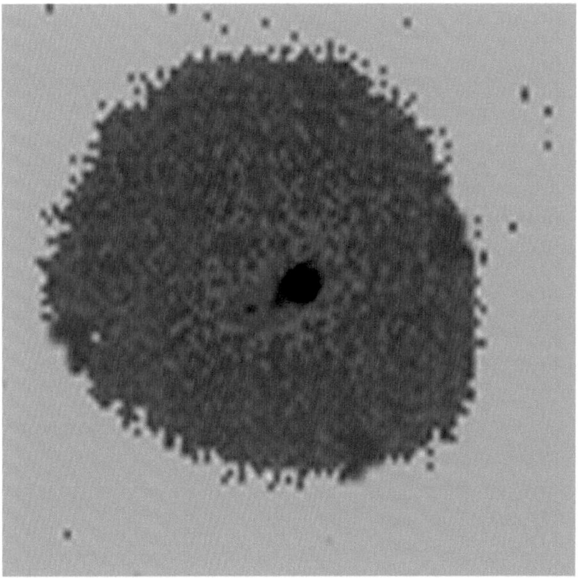

Fig. 2.6 Burn paper pattern. The *central dark spot* is the filament core (from Chin et al., 1999)

many experiments. We call this surrounding zone the background energy reservoir (Mlejnek et al., 1998, 1999; Kandidov et al., 2003a, b).

2.2 Experimental Definition of a Filament by Burn Paper

In the laboratory, it is easiest to use a burn paper to show the existence of a filament. This technique is at best ambiguous. This is because the damage threshold of the burn paper is low requiring normally the absorption of 3 to 4 photons whose intensity threshold is around 10^{10}–10^{11} W/cm^2 or less. However, ionizing air molecules takes about 8 photons and the intensity is clamped around 5×10^{13} W/cm^2. Thus, soon after self-focusing starts and well before ionization takes place, the burn paper already shows a burn mark on the surface. After the end of filamentation, the peak power is lower than the critical power for self-focusing. There will thus be no collapse of the pulse and there will be no more ionization. The filament core, now being a single fundamental mode because of self-spatial filtering (see later), starts to diffract out. In the first part of this propagation, the intensity is still high enough to self-focus but is overcome by diffraction in such a way that the divergence of the pulse is very small. This would still create an impact on the burn paper. Thus, the definition of a filament depends very much on the sensitivity of detection of intensity inside the propagation zone, if we use intensity as the criterion. We shall come back to this subject later more quantitatively when we describe the full evolution of a filament.

2.3 Single Filamentation Physics

We discuss single filamentation using air as the propagation medium. In order to describe the basic physics of filamentation (see for example, Boyd, 2003; Brodeur et al., 1997; Chin, 2006; Chin et al., 1999, 2005, 2008), we consider a short laser pulse, say, 50–100 fs in duration, from the most popular Ti-sapphire laser system (central wavelength 800 nm).

2.3.1 Slice-by-Slice Self-Focusing

For self-focusing to occur, the transverse spatial intensity distribution of the pulse across the wave front should not be uniform. We approximate the pulse as a plane-wave pulse and assume that the intensity distribution across the pulse's transverse cross section is Gaussian. We shall follow the propagation of the central (most powerful) "slice" of the pulse. The thickness of this "slice" is at least $c\tau$, where c is the speed of light in vacuum and τ is the period of oscillation of the electromagnetic wave. This is because we are talking about an intensity that is defined as the Poynting vector averaged over at least one cycle of oscillation. The propagation of this slice is similar to that of a wave front. If the intensity at the central zone of the slice is high enough so that the nonlinear Kerr effect cannot be neglected, the index of refraction of the central zone will be given by $n = n_0 + n_2 I$, while the index at the edge of the slice will be $n = n_0$. Here, n_0 is the linear index of refraction in air and $n_2 I$ is the Kerr nonlinear index of refraction; n_2 and I being the coefficient of the Kerr nonlinear index of refraction and the local intensity, respectively. The speed of propagation of the slice is given by c/n. Hence, the central part of the slice propagates slower than the rest of the slice, giving rise to a concave wave front as shown in Fig. 2.7. This is the beginning of self-focusing. However, this self-focusing effect is not sufficient to guarantee filamentation because there is always a linear diffraction of the pulse that will cause the pulse to diverge as it propagates further. If the self-focusing effect is not strong enough to counteract the diffraction effect, the consequence is a slowly divergent pulse, slower than that due to pure linear diffraction. Consequently, the pulse's diameter looks almost constant over some distance of propagation.

When the natural linear diffraction of the pulse is just balanced by self-focusing, the peak power equals the so-called critical power for self-focusing. Through a solution of Maxwell's equations for a nonparaxial CW Gaussian beam, the critical power for self-focusing is given by $P_c = \frac{3.77\lambda^2}{8\pi n_2 n_0}$, where λ is the central wavelength of the pulse (Marburger, 1975). This expression shows that the critical power for self-focusing depends only on n_2, n_0 and λ and is independent of the intensity. Thus, when the peak power of the pulse is higher than the critical power for self-focusing, the slice shown in Fig. 2.7 will continue to curve forward as the wave front propagates further. If the peak power is only very slightly higher than P_c, the group

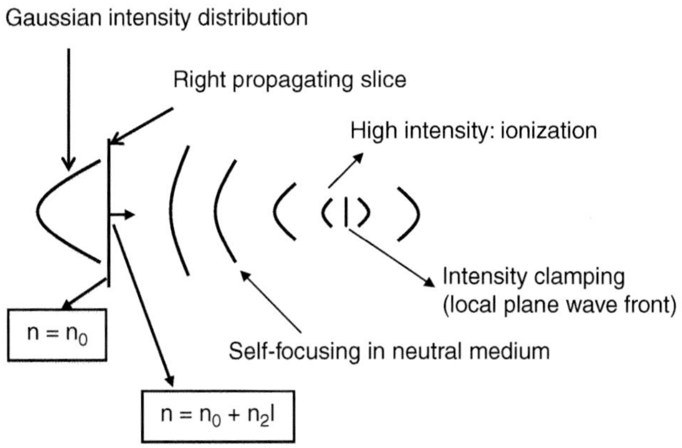

Fig. 2.7 Schematic illustration of the self-focusing of a slice of a femtosecond laser pulse in an optical medium. The initial plane-wave slice has a Gaussian intensity distribution across the transverse plane. The central part having a higher index of refraction due to Kerr nonlinear index increase, propagates slower than the rest of the slice, resulting in a concave slice front, which means focusing. When the self-focusing effect is stronger than linear diffraction and the effect of GVD, the radius of curvature of the slice will keep decreasing; hence, self-focusing becomes stronger and stronger. Soon, the intensity in the self-focusing zone becomes so strong that tunnel ionization starts to be significant. The resultant plasma slows down the focusing and balances it at the self-focal plane indicated by a *straight line*, where a maximum intensity is reached (intensity clamping) before the slice diverges out into the background reservoir

velocity dispersion (GVD) will lengthen the pulse after a short distance of propagation; this lowers the peak power to a value below P_c and the pulse will again diverge slowly through diffraction. However, with femtosecond laser pulses, it is easy to obtain a high peak power that can readily overcome both linear diffraction and GVD. A few tens of percent higher than P_c are enough (Brodeur and Chin, 1998, 1999). Once such self-focusing starts, it will not stop. Thus, the slice keeps curving into a smaller and smaller zone as it propagates while the intensity becomes higher and higher (Fig. 2.7). Soon, the high intensity in the self-focal zone will tunnel ionize (see Chapter 1 and Chin et al., 1985; Chin, 2004) air molecules, resulting in the generation of a weak plasma. The change in the index of refraction of the slice propagating in a plasma is[1] $(\Delta n)_p \cong -\frac{4\pi e^2 N_e(t)}{2m_e\omega_0^2}$, where N_e is the electron density,

---

[1]The contribution to the index of refraction by a plasma can be obtained in the following way. In a plasma, from any optics text book, by assuming that only the electrons contribute, the index of refraction is given by

$$n^2 = 1 - \omega_p^2/\omega^2 \qquad \text{(f2.1)}$$

where the subscript p denotes plasma; the plasma frequency $\omega_p = \left[4\pi e^2 N_e(t)/m\right]^{1/2}$ (where e and m are the electron charge and mass in cgs units, respectively and $N_e(t)$ is the MPI/TI generated time dependent density in cm^{-3}, i.e., $N_e(t)$ depends on the intensity of the laser). In air, $\omega_p << \omega$. This is always the case in the self-focus where the intensity is clamped down to between $10^{13}$and

e and m_e are the electronic charge and mass, respectively, and ω_0 is the central frequency of the pulse. The index of refraction of the central part of the slice is, thus $n = n_0 + n_2 I - \frac{4\pi e^2 N_e(t)}{2m_e\omega_0^2}$. This will increase the speed of propagation of the central part of the slice; i.e., the curvature of the slice starts to flatten out, but it is still focusing so long as $n_2 I > \frac{4\pi e^2 N_e(t)}{2m_e\omega_0^2}$. Thus, the intensity is still increasing. The electron density increases very rapidly with the intensity because tunnel ionization is a highly nonlinear process. We approximate such an increase as being governed by an effective power law according to an experimental observation (Talebpour et al., 1999a); i.e., $N_e(t) \propto I^m$, where m is the effective nonlinear order of ionization. In air, m is about 8 (Talebpour et al., 1999a). The effective index of refraction of the central part of the slice is thus $n = n_0 + n_2 I - \frac{4\pi e^2}{2m_e\omega_0^2} kI^m$, where k is a proportionality constant. Qualitatively, this means that the free electron term will quickly catch up with the Kerr term until they are equal; i.e., until $n_2 I = \frac{4\pi e^2}{2m_e\omega_0^2} kI^m$. At this point, Kerr self-focusing balances free electron defocusing, and the central part having now an index of refraction n_0 propagates at the same speed as the rest of the slice. There is no more focusing (i.e., a local plane wave front, see Fig. 2.7) and the intensity is highest at this balancing point. This is the condition of intensity clamping (Kasparian et al., 2000a; Becker et al., 2001; Liu et al., 2002) because further propagation would lead to an index at the central part smaller than n_0. The slice will start to diverge. That is to say, during self-focusing of a powerful femtosecond laser pulse in an optical medium, there is a maximum intensity that self-focusing can reach. In air, it is around 5×10^{13} W/cm^2 (Becker et al., 2001). The energy in the defocusing slice will be reduced a little due to the loss in ionization. After passing through the self-focus, the central slice is returned (defocused) back to the remaining part of the whole pulse or to the background reservoir (Mlejnek et al., 1998, 1999; Kandidov et al., 2003a). This background reservoir is an important concept in considering the physics of filamentation. An experimental and numerical study of the background reservoir is given in (Liu et al., 2005b) and will be discussed later in the book.

10^{14} W/cm^3 (Théberge et al., 2006) at which single ionization dominates. The electron density in a filament in air generated by a 50 fs/800 nm laser pulse has been measured to be of the order of 10^{16}/cm^3. This gives $\upsilon_p = \omega_p/2\pi = 3 \times 10^9$ Hz which is much smaller than the optical frequency ($\sim 10^{14}$ Hz). Hence, Eq. (f2.1) becomes

$$n = \left[1 - \omega_p^2/\omega^2\right]^{1/2} \cong 1 - \omega_p^2/2\omega^2 \tag{f2.2}$$

When $N_e(t) = 0$, i.e., in vacuum, $n = 1$. That is to say, $n = 1 + \Delta n_{\text{plasma}}$ where Δn_{plasma} is the contribution of the plasma to the index of refraction.

$$\Delta n_p \cong -\omega_p^2/2\omega^2 = -\frac{4\pi e^2 N_e(t)}{2m\omega^2} \tag{f2.3}$$

which is negative.

The critical power for self-focusing, which is inversely proportional to the coefficient of the nonlinear index of refraction, n_2, is not constant in air. It depends on the response of the medium to the pulse duration. We note that the response of a medium to an electromagnetic wave is essentially the induced polarization (dipole moment per unit volume). When the pulse duration is shorter than 100 fs, only an "instantaneous" electronic response (induced polarization due to a pure electronic oscillation that can follow the field) is fast enough to contribute to the total (linear and nonlinear) polarization, which in turn contributes to the total index of refraction and to n_2. When the pulse is longer so that the interaction time is longer, both the electronic and the nuclear responses involving the Raman transition (excitation of a molecular vibration) contribute to a larger value of n_2. This lowers the critical power. In air, our recent experimental measurement (Liu and Chin, 2005) shows that P_c changes from about 10 GW for pulse durations shorter than 100 fs to about 3 GW for pulses longer than 100 fs.

The plasma density in the self-focus depends on the external condition. For free propagating beam in one atmospheric air, it is measured to be around 10^{14} cm^{-3} (Théberge et al., 2006a). This value increases when using an external focusing lens. The shorter the focal length is, the higher the density will be. The highest measured value is of the order of 10^{18} cm^{-3} using a 10 cm focal length lens in air (Théberge et al., 2006a).

The central slice will self-focus at a self-focusing distance z_f from the beginning of the propagation in the medium given by[2] (Marburger, 1975)

$$z_f = \frac{0.367ka_0^2}{\left\{\left[\left(P/P_c\right)^{1/2} - 0.852\right]^2 - 0.0219\right\}^{1/2}} \qquad (2.1)$$

where k is the wave number, a_0, the radius of the beam profile at 1/e level of intensity, and P, the peak power of the slice. The slice in front of the central slice will then self-focus at a later position in the propagation direction according to Eq. (2.1) because its peak power is lower than that of the central slice. It will undergo the same processes, namely, self-focusing, intensity clamping, and de-focusing, and will return the (slightly lowered) energy back to the background reservoir, and so on for successive front slices whose peak powers are higher than the critical power (Figs. 2.8 and 2.9). Thus, the front part of the pulse will become thinner and thinner as the

[2]Equation (2.1) is the solution of the nonlinear Schroedinger equation coming from the Maxwell's equations without GVD. The initial conditions are: (1) the laser beam is a continuous one (CW); (2) it is a paraxial cylindrical beam with a spatial Gaussian distribution of intensity at the input of the medium; (3) slowly varying envelope approximation is used. During the propagation, the beam is deformed and numerical technique is used to obtain Eq. (2.1). Interestingly, this equation was found to be applicable even down to pulse duration of about 10 cycles of oscillation of the field. For example, in the case of a Ti-sapphire laser at the wavelength of around 800 nm with a pulse duration of the order of 100 fs, this equation was found to describe well the beginning of filamentation (Brodeur et al., 1997).

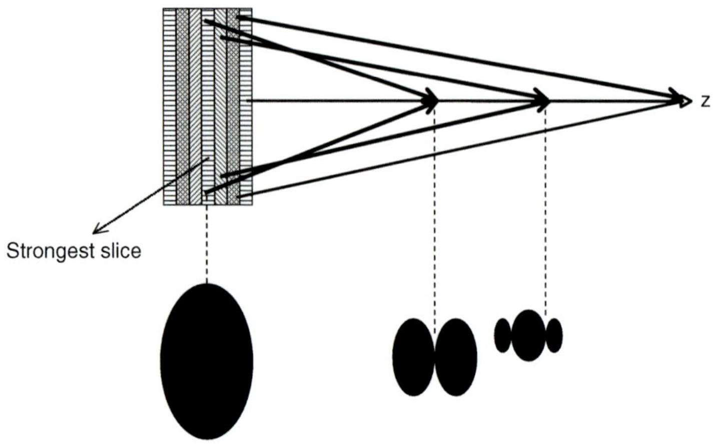

Fig. 2.8 Slice-by-slice self-focusing in an ideal case if there is no ionization at the self-focus. The *top part* shows the pulse represented by a distribution of slices. The *central* most powerful slice self-focuses at a shorter distance in the direction of propagation z. The weaker slices at the side self-focus at longer distances. The *lower part* shows an ellipsoid that represents the pulse. The vertical dimension is power and the horizontal dimension is spatial length of the pulse. The ellipsoid is squeezed down as it propagates into successive structures shown at the *right side* of the ellipsoid. However, in reality, ionization takes place at the self-focus. The *front part* of the pulse (ellipsoid) will still undergo the same squeezing, while the *back part* will squeeze into the ionization zone of the preceding slice. This will induce interaction with the plasma. The consequence is shown in Fig. 2.9

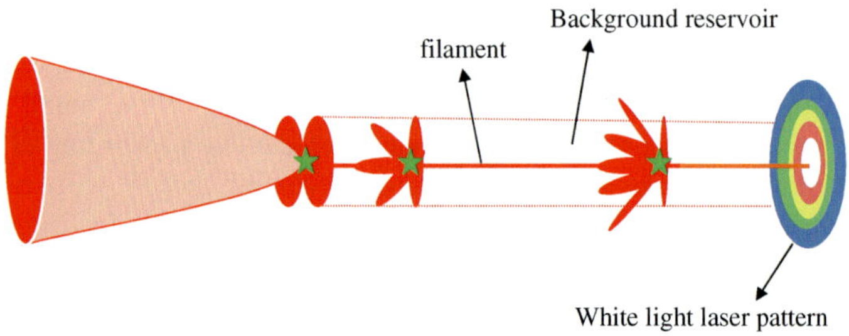

Fig. 2.9 A schematic diagram of the evolution of a femtosecond laser pulse propagating in an optical medium. The direction of propagation is towards the *right hand side*. The pulse is represented by the ellipse at the *left*. The width of the ellipse represents the pulse's spatial width (c times the pulse length). This width is of the order of 30 μm for a 100 fs pulse; hence, a sheet of light. As such, the ellipse could be imagined as a "pancake" of light with a lot of photons inside. The *central slice* of the pulse self-focuses to a small area where the resulting high intensity ionizes the air molecules (*star*). The *front part* keeps on self-focusing, becoming thinner and thinner. The *back part* encounters the plasma left behind by the *front part* and undergoes self-phase modulation; the resultant intensity (*field*) distribution becomes non-uniform and is represented by the diverging "splashes" of elongated "sub-pancakes" of light. At the end of the propagation, the pulse degenerates into a colorful *white* light laser pulse (from Chin et al., 2005)

pulse propagates. The back slices symmetrical to the front slices will, in principle, also self-focus at a position slightly behind the self-foci of the front slices (Fig. 2.8). However, this will never happen because it will encounter the plasma left behind by the central and successive front slices. These back slices will thus self-focus into and interact with the plasma giving rise to a complex intensity distribution (see for example, Kandidov et al., 2003a; Aközbek et al., 2000). In general, the energy in the back part of the pulse will still be confined inside the highly deformed body of the pulse or the background reservoir (Fig. 2.9).

During the propagation, repeated processes of Kerr self-focusing in the neutral gas and self-defocusing in the self-generated weak plasma of the slices in the front part of the pulse result in a continuous series of hot spots along the propagation axis. This gives rise to the perception of a filament, and hence, filamentation (Brodeur et al., 1997; Kosareva et al., 1997a; Mlejnek et al., 1998; Chiron et al., 1999) (Fig. 2.9). Since the energy loss in the ionization process is small, the pulse can repeat the whole process again, resulting in what we call self refocusing (Mlejnek et al., 1998; Talebpour et al., 1999b). This is manifested by successive sections of brighter lines in Fig. 2.5.

Before ending this section, the author would like to comment upon the so-called Marburger formula; i.e., Eq. (2.1). Very often, one says that z_f is independent of intensity because the formula seems to indicate that z_f is not a function of the intensity of the pulse (slice). However, a closer look reveals an implicit dependence on the intensity through the radius a_0. This is the radius of the intensity distribution of a pulse with spatial Gaussian distribution. It is related to the intensity through $I_p = P_p/\pi a_0{}^2$ where P_p and I_p are the peak power and peak intensity, respectively. Thus, for a constant pulsed energy with a constant pulse duration (i.e., a constant peak power), changing the radius automatically changes the peak intensity or vice versa. This in turn changes the value of z_f, i.e., z_f also depends on the intensity.

2.3.2 Intensity Clamping

Intensity clamping is a profound physical manifestation of self-focusing and filamentation. It sets an upper limit to the intensity at the self-focus not only in air but also in all optical media. Even if one tries to focus the pulse, so long as the focal length is not too short (Liu et al., 2003), self-focusing will always start before the geometrical focus. Thus, the intensity at the geometrical focus is either lower than, or as high as that inside the self-focal zone in air. The consequence of this intensity clamping is far reaching. In air, one can have self-focusing at a long distance but one cannot further increase the intensity inside the self-focus, not even by significantly increasing the energy of the pulse to many times the critical power. In practice, there will only be an increase in the number of self-foci (multiple filamentation), each of which will have similar peak intensity. The "dream" of reaching an enormous intensity (that might induce a nuclear reaction, for example) on remote targets in the atmosphere has to be forgotten in the current context. On the other hand, if the beam profile is so smooth that only a single filament will persist while the peak power is increased significantly to many times the critical power for self-focusing, the diameter, and hence the volume of the filament will increase while the intensity inside this larger volume will still be clamped (Théberge et al., 2007a). Also, multiple re-focusing will take place. In practice, this is a tough condition to fulfill because any little fluctuation in intensity on the beam profile will lead to local self-focusing so long as the local power is higher than the critical power for self-focusing. This again results in multiple filaments. Furthermore, because the intensity is almost constant, any interaction making use of, or sampling the filament core, will result in a very stable outcome. One example is third harmonic generation (Aközbek et al., 2002).

The clamped intensity in air (or gases) is independent of pressure. Thus, when filamentation occurs at a high altitude in the atmosphere, the clamped intensity is always the same as that at sea level because when intensity clamping occurs, the nonlinear Kerr index change and the index due to plasma generation are equal: $n_2 I = \frac{4\pi e^2 N_e(t)}{2m_e\omega_0^2}$. Both n_2 and $N_e(t)$ are linearly proportional to the gas density since $N_e(t)$ comes from tunnel ionization of the individual molecules. Hence, the gas density cancels out on the two sides of the equation, leaving behind an equation for the solution of the same clamped intensity I at any pressure. This claim is verified experimentally by the author and collaborators (Bernhardt et al., 2008) using He as the target gas because of its simplicity and unambiguity.

One might ask if GVD and linear diffraction could also have led to intensity clamping since they both would be able to stop self-focusing from developing. In linear diffraction, if the peak power is smaller than the critical power, there will never be a strong focal spot in the propagation. The beam will keep on growing in diameter. When the peak power equals the critical power, the beam will self-focus at infinity. (This is the definition of the critical power.) If the peak power is increased further, self-focusing overcomes linear diffraction and will self-focus towards a singular point. If nothing else happens, ionization will occur and the intensity is clamped. However, if GVD is large; i.e., if the bandwidth of the pulse is large, different frequencies in the pulse will propagate at different group velocities leading to an elongation of the pulse. Consequently, if the initial peak power is not high enough, it could become lower than the critical power. One could say that at the point where GVD and linear diffraction balance self-focusing, the intensity is clamped. But this balancing intensity depends on the bandwidth of the pulse (which can be changed at will, in principle) and will not lead to the occurrence of a series of hot spots (filament). Thus, we could not use this balancing point as the criterion of intensity clamping. When both linear diffraction and GVD are overcome by self-focusing, the beam will keep on focusing to a small spot until ionization occurs and balances self-focusing. It is this balancing point that is unique because it depends on the ionization potential (probability) of the molecules/atoms in the medium. The ionization potential of an atom or a molecule cannot be changed. It is fixed by nature rendering the clamped intensity unique.

2.3.3 *Is There Optical Breakdown During Filamentation?*

Very often, when the idea of ionization at the self-focus is proposed to be the mechanism that balances the effect of self-focusing resulting in intensity clamping, the "instinctive" reflex of many experienced scientists would be that the ionization is naturally an optical breakdown. This is because it involves (self) focusing a strong laser pulse into a transparent optical medium. This would lead to the belief that the plasma generated by optical breakdown would stop further propagation of the laser pulse, as is the case for nanosecond and picosecond pulses (see Chapter 1). The peak laser power involved in such experiments can be a few MW (Chin and Isenor, 1967 and references therein). Such a phenomenon is shown in Fig. 2.10. The 10 ns second harmonic (532 nm) of a YAG-laser pulse is focused by a 50 cm focal length lens in air. The plasma grows towards the lens and is stronger and longer in front of the focus than behind. This is because once the optical breakdown threshold is reached during the rising part of the pulse, a plasma is created at the geometrical focus (see Fig. 2.11a). There is a lot of time for this plasma to absorb the rest of the laser radiation and grow and scatter the remaining part of the pulse. The self-focusing threshold is much higher than the breakdown threshold and hence will never be reached before the strong growth of the plasma. The consequence is that the length of the plasma cannot be very long. Thus, during self-focusing of an fs laser pulse, if the above mechanism did take place, we would never have observed a long filament. Since a long filament did occur, the ionization model should not be valid.

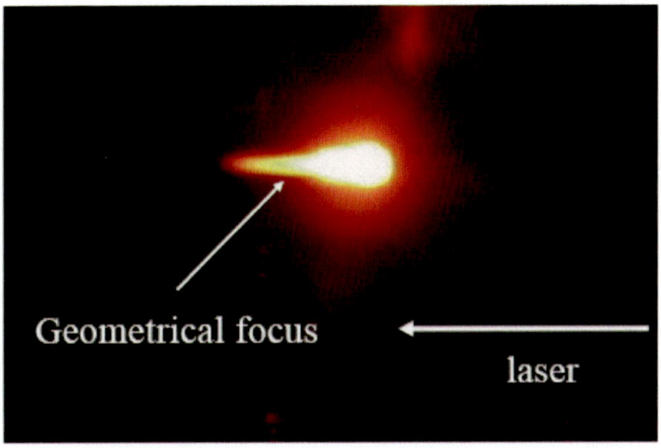

Fig. 2.10 Breakdown by focusing a 700 mJ/10 ns laser pulse at 532 nm in air by a 50 cm focal length lens. The plasma grows towards the laser while not much laser energy penetrated through the plasma, and the plasma is much weaker at and behind the geometrical focus. The total length of the plasma is of the order of 20 mm (picture taken by S. Mehdi Sharifi in the author's laboratory)

As discussed in Chapter 1, the mean free time of a free electron is about 1 ps in atmospheric air, while usually filamentation experiments are done using pulses of a few tens of fs or longer. There is thus not enough time for the free electron to undergo collision with the surrounding molecules before the laser pulse is over.

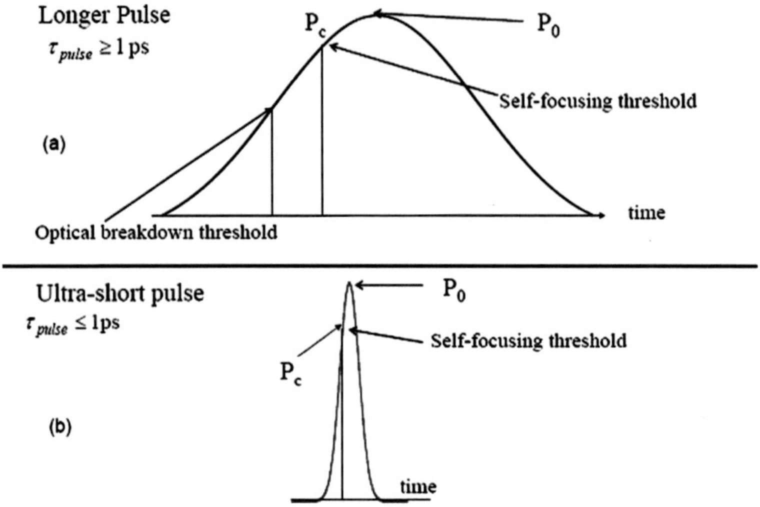

Fig. 2.11 (**a**) Long laser pulse induced optical breakdown in air (gas); (**b**) femtosecond laser induced tunnel ionization (no breakdown) in air (gas). P_c is the critical power for self-focusing, P_0 the peak power of the laser pulse

Hence, even a first round of inverse Bremsstrahlung is not possible in atmospheric air. This is shown in Fig. 2.11b. The sharp rise of the pulse is such that well before the threshold of breakdown is reached, self-focusing takes place. After that, the plasma generation in the self-focal region is mainly direct tunnel ionization of the molecules.

Also, the slice-by-slice self-focusing mechanism does not allow the full temporal (spatial) part of the fs laser pulse to focus into the same focal volume. Each slice self-focuses at a difference zone in space. This slice is thus shorter than the full pulse width. The interaction time for ionization is thus shorter than the full pulse duration resulting in less ionization or the generation of less plasma. Moreover, since the pulse undergoes self-compression in time during filamentation, the interaction time would be of the order of one or two cycles of field oscillation. New interaction physics is expected.

In condensed matter, because the mean free time is about 1 fs, there would be sufficient time to induce breakdown. However, because of the slice-by-slice self-focusing mechanism, the interaction time inside a slice is of the order of one or two cycles of field oscillation (see Section 2.3.1). This would mean that there is still not much time for the plasma to grow through full cascade ionization. The plasma density inside the filament in a condensed matter is thus in the order of 10^{18}–10^{19} cm^{-3} which is 3–4 orders of magnitude lower than the density of condensed matter. This is a major reason why femtosecond laser pulses create much less damage than nanosecond and picosecond laser pulses in optical media.

2.3.4 Effect of External Focusing

In many experiments, an external lens/mirror is used to force self-focusing within the limited length of the medium. Self-focusing is then reinforced by the external lens such that the self focus appears before the position of the geometrical focus. The new self-focusing distance z_f' satisfies the lens transformation equation: $1/z_f + 1/f = 1/z_f'$, where z_f is the self-focusing distance given by Eq. (2.1) and f is the geometrical focal length of the lens/mirror (Talanov, 1970). Figure 2.12 (Liu et al., 2003) shows the measured fluorescence from nitrogen molecules from inside the filament regions using three different focal lengths. For long geometrical focal length, filamentation occurs before the geometrical focus (Fig. 2.12a) while the intensity at the geometrical focus is low. In such cases, the filament length is roughly the distance between the self-focusing position at z_f' and the geometrical focus. When the geometrical focal length becomes shorter, the filament extends into the geometrical focus (Fig. 2.12b). Here, the effect of geometrical focusing becomes evident. At the geometrical focus, the stronger external focusing generates a plasma stronger than that generated by the filament. Self-focusing and geometrical

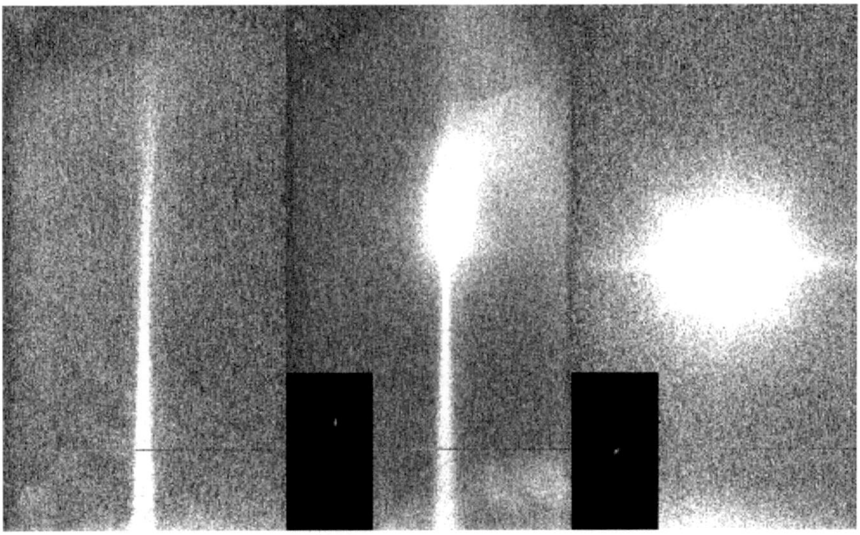

Fig. 2.12 The "competition" between self-focusing and geometrical focusing in air. *Images* show the measured fluorescence from nitrogen molecules from inside the weak plasma regions. They are taken via a grating's zeroth order image; the 42 fs laser pulse propagates towards the top; input energy, 5 mJ/pulse. (**a**) $f = 100$ cm, (**b**) $f = 30$ cm, (**c**) $f = 5$ cm; *inset*: low contrast pictures of $f = 30$ cm and $f = 5$ cm. When the geometrical focal length is long, filamentation occurs before the geometrical focus (**a**). When the geometrical focal length becomes shorter, the filament ends into the geometrical focus (**b**). Self-focusing and geometrical focusing become indistinguishable when the geometrical focal length is very short (**c**) (from Liu et al., 2003)

focusing become indistinguishable when the geometrical focal length is very short (Fig. 2.12c).

Intensity clamping still occurs with geometrical focusing. However, the intensity would increase slightly while the plasma density would increase drastically. Qualitatively, when geometrical focusing is added to self-focusing, it is effectively similar to having a stronger focusing; i.e., a concave wave front with a larger curvature as compared to that of a self-focusing plane wave front. In order to overcome this stronger focusing, the self-generated plasma has to become denser so that the negative lensing effect can balance the focusing effect. To see more quantitatively how geometrical focusing influences the plasma density at the focus where intensity clamping occurs, we go back to Sections 2.3.1 and 2.3.2 as well as Fig. 2.7. The change in the index of refraction in the lens along the propagation axis with respect to the edge of the lens is Δn_{lens}. This should be added to the nonlinear increase of the index of refraction $\Delta n_{\text{Kerr}} = n_2 I$ and together, they are balanced by the self-generated plasma's index of refraction (Chen, 2008, private communication) which is

$$(\Delta n)_p \cong -\frac{4\pi e^2 N_e(t)}{2m_e\omega_0^2}$$

i.e.,

$$\Delta n_{\text{lens}} + n_2 I \cong \frac{4\pi e^2 N_e(t)}{2m_e\omega_0^2}$$

It is evident that to reach this balance, $N_e(t)$ must be larger than when there is no geometrical focusing (free propagation); i.e., the intensity must be higher when an external geometrical focusing is used. Experimentally, it was observed by Théberge et al. (2006a). However, the increase in intensity is not very much even though the plasma density changes by two to three orders of magnitude. This is because the plasma is generated through the highly nonlinear multiphoton/tunnel ionization process. A little change in intensity would lead to a huge change in the plasma density.

2.3.5 Background Energy Reservoir

One important consequence of slice-by-slice self-focusing is the concept of the so-called background energy reservoir or energy reservoir or simply reservoir. A self-focusing slice will generate a plasma which defocuses the slice. The energy of this slice is thus not always confined along the propagation axis. It goes back to the surrounding region but is always a part of the pulse. After a round of slice-by-slice self-focusing of the whole pulse, a plasma column along the axis is left behind (filament).

Because the so-called filament or the series of self-foci are fed from outside the filament core, one has to be careful in using different words to describe the same phenomenon. For example, self-channeling is used very often to describe filamentation. But the word "self-channeling" might implicitly be erroneously interpreted by the non-initiated to mean that it is the whole pulse that self-focuses and continues channeling its way through the medium as if it were a "light-bullet". The latter name is also used from time to time for reasons of sensation. But in fact they all mean the same physics of slice-by-slice self-focusing.

The existence of a wide background reservoir has been noticed already in the first demonstrations of filamentation in air (Braun et al., 1995; Nibbering et al., 1996; Brodeur et al., 1997). It has been interpreted by results of computer simulations as an energy reservoir of the filament core (Mlejnek et al., 1998). Several experiments (Brodeur et al., 1997; Dubietis et al., 2004b; Liu et al., 2005a, b) showing that filaments are terminated after blocking the outer part of the pulse except its core, provided support for this interpretation. Recently, it has been further shown in an experiment (Courvoisier et al., 2003) and subsequent numerical simulations (Dubietis et al., 2004b; Kolesik and Moloney, 2004; Skupin et al., 2004) that filaments are robust after a collision with a water droplet. This observation has been attributed as being due to an energy transfer from the background to the filament core. Also, results of a variational analysis (Arevalo and Becker, 2005) indicate that already the process of self-focusing is strongly influenced by the wide background of the pulse. The following gives a detailed experiment that shows the existence of the background energy reservoir. The size of the reservoir and the portion of the pulse energy located in the background are also estimated (Liu et al., 2005b).

A Ti-sapphire laser was used. It gave out 800 nm/50 fs pulses with a spectral bandwidth of 23 nm at FWHM, a repetition rate of 10 Hz and energy of 2.5 mJ/pulse. The peak power is many times larger than the critical power of self-focusing in air (~6–10 GW) (Liu and Chin, 2005). The initial beam diameter has been reduced to 2 mm at FWHM by a telescope consisting of a pair of a plano-convex lens ($f = 40$ cm) and a bi-concave lens ($f = -20$ cm) (Fig. 2.13). Using this set-up, a long single filament was obtained in air.

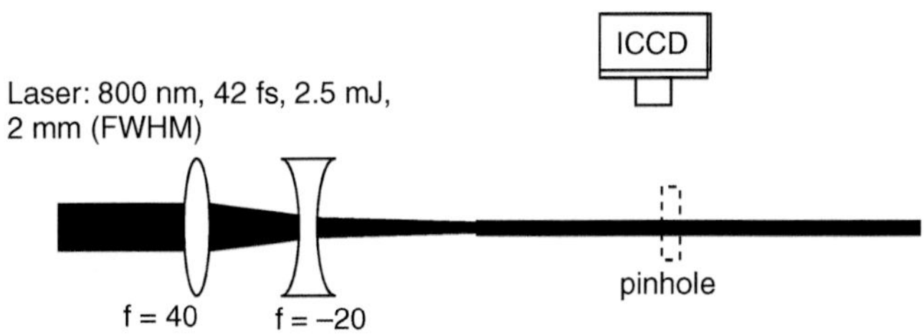

Fig. 2.13 Experimental set up to demonstrate the existence of background energy reservoir

Pinholes of different diameters (220 μm to 2 mm) were introduced coaxially to the propagation axis of the beam. The evolution of the filament formation was observed by using an intensified CCD camera (ICCD, Princeton Instruments, PI-MAX 512) to image the fluorescence emitted from the nitrogen molecule (N_2) and nitrogen ion (N_2^+) in the filament core (Talebpour et al., 2001). The ICCD camera was installed perpendicularly to the pulse propagation axis and the fluorescence signal was collected and imaged onto the ICCD detector by using a single plano-convex lens made of fused silica with a focal length of 63 mm and a diameter of 38.1 mm. A band-pass filter (1 mm-thick UG11, Corion) together with a 0° incident 800 nm dielectric mirror were placed in front of the camera to integrate the light emission over the strongest N_2 and N_2^+ bands around 350 nm while rejecting the scattered light from the pump laser (Hosseini et al., 2003). There is a good overlap between the band–pass filter transmission curve and the strongest N_2 and N_2^+ bands around 350 nm. With this configuration, about 0.5 m along the laser propagation axis was covered by the field of view of the ICCD detection system. Furthermore, a slim black screen (~ 3 cm wide) was put beside the pinhole to avoid strong scattering of the laser light from the metal surface of the pinhole to the detector.

Images acquired by the ICCD camera are shown in Fig. 2.14a–e. Each image results from an accumulation of 1,000 laser shots. The laser propagates from the left to the right and the propagation distance is scaled with respect to the concave lens of the telescope. Figure 2.14a represents the free propagation of the pulse (without pinhole). A bright line over almost half a meter is seen along the propagation axis. The onset of strong nitrogen fluorescence, and hence, the beginning of the filament is located near the left side of the image. The signal decays toward the right side of the figure after it has reached its maximum at a distance of about 1.7 m. The other panels on the left hand side of Fig. 2.14 show the images for pinholes of different diameters inserted at 1.73 m, starting with a diameter of 220 μm (Fig. 2.14b) and up to 2 mm (Fig. 2.14e). The black gap between 1.7 and 1.8 m in these four panels appears due to the slim black screen mentioned above. It is seen that for the smallest diameter

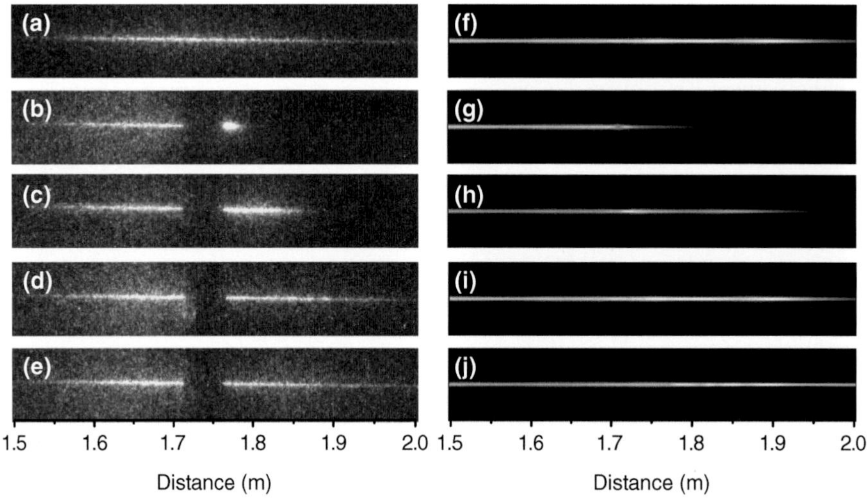

Fig. 2.14 *Left column*: Images of the nitrogen fluorescence signal recorded by an ICCD camera; *right column*: Electron density distribution from numerical simulations. (**a**, **f**): free propagation, (**b**, **g**): pinhole diameter 220 μm, (**c**, **h**): pinhole diameter 440 μm, (**d**, **i**): pinhole diameter 1 mm, (**e**, **j**): pinhole diameter 2 mm

(Fig. 2.14b, 220 μm), the plasma column is terminated by the pinhole (the white spot results from diffraction by the pinhole). In this case, no significant damage of the pinhole was found after the experiment which indicates the high stability of the laser system. That ensures the successes of our experiments. As the diameter of the pinhole is doubled (Fig. 2.14c, 440 μm), the filament partially survives up to about 1.9 m. When the pinhole diameter is increased to 1 mm (Fig. 2.14d) or 2 mm (Fig. 2.14e), the filament formation seems to be unchanged compared to the case of free propagation within the field of view of the camera.

Numerical simulations of the amplitude envelope A based on the nonlinear wave equation written in the retarded coordinate system (using the slowly varying envelope approximation) was performed. (The theory will be discussed later in the book. Readers not yet familiar with the theory can skip this part.)

$$2ik_0\frac{\partial A}{\partial z} = \Delta_{\perp}A - k_2k_0\frac{\partial^2 A}{\partial t^2} + 2\frac{k_0^2}{n_0}(\Delta n_{\mathrm{Kerr}} + \Delta n_{\mathrm{plasma}})A - ik\alpha A \tag{2.2}$$

In Eq. (2.2), diffraction, dispersion, and Kerr effect as well as plasma generation and energy losses due to multiphoton/tunnel ionization are considered. Values for k_0, k_2, Δn_{Kerr}, $\Delta n_{\mathrm{plasma}}$ and α are adapted from Kandidov et al. (2003a) and parameters of the laser system (2.5 mJ, 40 fs, 2 mm diameter at FWHM) have been used as initial conditions in the simulations. The effects of the pinholes are simulated by applying an energy transmission function at 1.7 m. The transmission function is set to be unity from the axis to the corresponding pinhole radius and smoothed toward the outer edge by a Gaussian function with 10 μm ($1/e^2$) width.

Electron density distributions obtained from the simulations are shown in Fig. 2.14f–j. The reason why the fluorescence is compared to the electron density is the following. The nitrogen fluorescence is the result of multiphoton/tunnel ionization of nitrogen molecules. Thus, the electron density and the fluorescence are related to each other directly. Each panel in Fig. 2.14f–j corresponds to the experimental result in the same row. It is seen that the experimental results are well reproduced by the simulations. Also, for the percentage of initial energy transmitted through the pinhole (Fig. 2.15a), there is a good agreement between the experimental results (open triangles) and those from simulations (solid squares) for all pinhole diameters. The results for the on-axis electron density distribution from numerical simulation, shown in Fig. 2.15b, reveal differences in the long-scale propagation outside the view of the ICCD camera between the cases of the 1 mm pinhole and the 2 mm pinhole. The refocusing peak, which appears roughly between 2.2 and 3 m for free propagation (solid line), is suppressed for the smaller pinhole (1 mm, dashed line) but not for the pinhole of 2 mm in diameter (dotted line). Thus, a background whose transverse dimension is about 5–10 times as large as the filament core and containing up to 50% of the pulse energy (see Fig. 2.15a), has to propagate together with the filament core in order to maintain the full length of the filament including one refocusing.

The results suggest that a diffraction of energy at the edges of the background lead to a leakage of energy which results in a termination of the filament. This

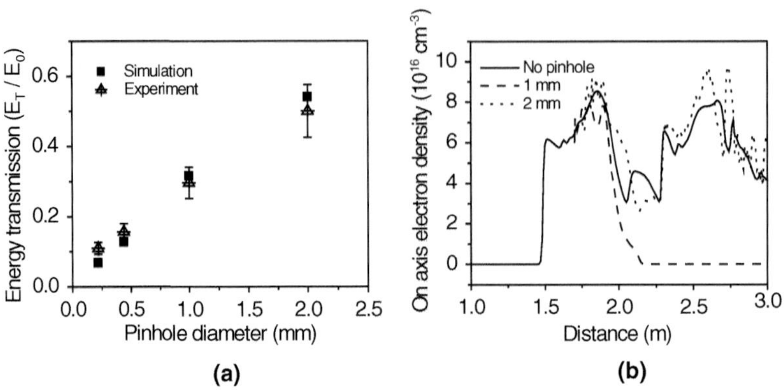

Fig. 2.15 (**a**) Energy transmission through the pinhole; (**b**) on axis electron density distribution from numerical simulations

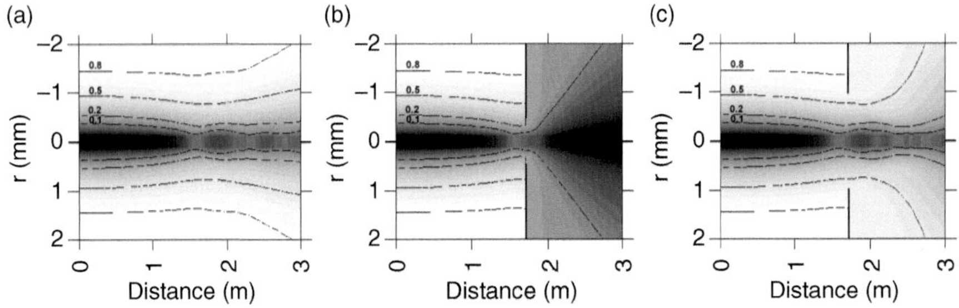

Fig. 2.16 Results of numerical simulations for the radial energy distribution as a function of propagation distance: (**a**) free propagation, (**b**) pinhole diameter 1 mm, (**c**) pinhole diameter 2 mm

interpretation is supported by the energy distributions obtained from the numerical simulation displayed in Fig. 2.16 for the cases of (a) free propagation, (b) with a pinhole of 1 mm in diameter and (c) with a pinhole of 2 mm in diameter. The horizontal and vertical axes correspond to the propagation distance and the radius respectively. Grey levels represent the energy percentage within a given radius r_0 about the propagation axis; the lighter the color, the larger the portion of energy enclosed. Four contour lines indicating the 10, 20, 50, and 80% levels are plotted to guide the eye. The bending of the line towards the propagation indicates a flow of energy towards the filament core and vice versa. The oscillations of the inner contour lines indicate the energy exchange between the filament core and the outer background (see Fig. 16a). Moreover, a large part of the pulse energy propagates together with the filament core before the energy starts to diffract out at the end of the filament. The pinholes initiate the diffraction at the edges of the background at an earlier stage of the filament formation compared to the case of free propagation (see Fig. 2.16b, c). For the smaller pinhole, this results in the termination of the filament before refocusing takes place.

The above results have important implications for long-range filament formation. It is obvious that for the maintenance of the high-intensity filament core which contains about 10% of the pulse energy only, the propagation of a wide background, in which 50% or more of the pulse energy is located together with the core, is needed. For the filament formation it is obviously more critical to avoid a diffraction of energy at the edges of the background than a collision with a (small) droplet near the center. While the former scenario appears to lead to a continuous leakage of energy towards the outside resulting in the termination of the filament sooner or later, the latter may just alter the energy flow within the core and the background. We may finally note that our results clearly rule out any self-guiding model for the filament core but do favor the moving focus (Brodeur et al., 1997) and spatial replenishment (Mlejnek et al., 1998) models. They may however raise the question if the whole filament, i.e., core *and* background, could be described by spatial soliton solutions or a superposition of such solutions (Dubietis et al., 2004b; Courvoisier et al., 2003).

The above shows that the formation of robust filaments is due to the wide low-intensity background around the tiny high-intensity core. We have shown that this background is more than 5 times larger than the filament core and contains up to 50% of the pulse energy. The robustness of the filament formation does depend crucially on diffraction of energy at the edges of the weak background.

2.3.6 Self-Spatial Mode Filtering

In a practical experiment, it is very rare, if not impossible, for the spatial mode quality of an intense fs laser pulse to be perfect. Thus, during filamentation experiments, the transverse spatial modes of the laser pulse "compete" with one another to form a self focus and it is the lowest order spatial mode that dominates. That is to say, only the lowest order mode inside the pulse will self-focus into a filament. All other higher order modes remain in the reservoir. This is called self-spatial filtering (Liu and Chin, 2007; Chin et al., 2007; Akturk et al., 2007).

We can understand the above phenomenon of self-spatial filtering as follows. When self-focusing starts, all the modes in a pulse will self-focus initially. However, since the lowest order spatial mode has a smallest diameter as compared to the other higher order modes, it will reach the self-focus first according to Eq. (2.1). At and near the self-focus of the lowest order mode, ionization takes place generating a plasma. All the higher order modes while still self-focusing, will encounter the plasma and be diffracted out into the background reservoir before forming self-foci. The result is that the central filament core comes from the self-focusing of a very clean fundamental mode whose divergence is smaller than the original beam divergence.

This self-spatial filtering would have profound impacts on many applications of the filament. One example is the generation of other pulses at other wavelengths through four-wave-mixing inside the filament (Théberge et al., 2006b). The mode of the newly generated pulse is a perfect lowest order mode because four-wave-mixing

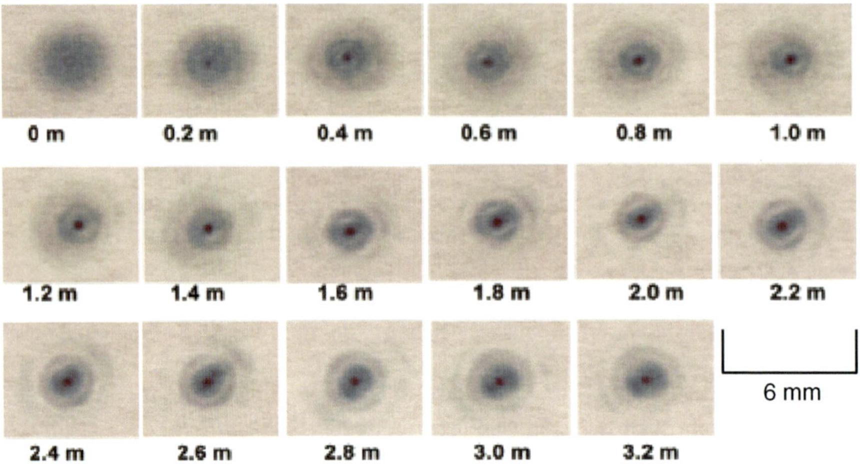

Fig. 2.17 Evolution of beam pattern during filamentation of a 2 mJ/45 fs Ti-sapphire laser pulse at around 800 nm. The parallel beam was created by using a telescope (two lenses) to reduce the beam diameter down to 2 mm diameter (FWHM). The pictures were the patterns scattered from a *white* piece of paper intercepting the filament at different positions. Zero means next to the exit of the telescope. The CCD camera measured only the image at 800 nm wavelength from a smallest possible angle with respect to the laser propagation axis (about 15°)

is a nonlinear process interacting with the high intensity core without implicating the low power reservoir practically. Thus, only the fluence distribution in the core is reflected in the newly generated pulse (Théberge 2006a). Another manifestation of the self-spatial filtering shown in Fig. 2.17 shows the evolution of the beam patterns during filamentation. The central symmetrical hot spot persists all along the filament.

2.3.7 *Self-Phase Modulation, Self-Steepening and White Light Laser (Supercontinuum)*

Because of the shortness of the femtosecond laser pulse, one of the consequences of nonlinear interaction with the medium during filamentation is self-phase modulation. The interaction of the high intensity in the self-focal regions with the neutral gas and with the plasma results in the modulation of the phase of the pulse; i.e., self-phase modulation (SPM) (See for example, Boyd, 2003). The consequence is spectral broadening towards both the red and the blue sides. SPM in an optical medium is caused by the temporal variation of the refractive index.

It is understandable that (also according to theoretical calculation) at the self-focus, where self-focusing is balanced by self-defocusing due to the plasma generation, the wave front becomes plane. To simplify the explanation, we assume that the self-focused wave is a plane wave. In this plane wave approximation, the plane wave front at the self-focus is given by the function.

$$F(z,t) = \exp\{i[\omega_0 t - kz]\} = \exp\left\{i\left[\omega_0 t - \frac{\omega_0 n}{c}z\right]\right\} \tag{2.3}$$

where z is the propagation distance, ω_0, the central angular frequency of the laser.

$$n \cong n_0 + \Delta n\,(t) \tag{2.4}$$

$$\Delta n\,(t) = n_2 I(t) - \frac{4\pi e^2 N_e(t)}{2m_e\omega_0^2} \tag{2.5}$$

Here, $n_2 I(t)$ is the Kerr nonlinear refractive index of the neutral gas, $I(t)$ being the intensity. The last term is the plasma contribution (see footnote 1 of Section 2.3.1) where $N_e(t)$ is the electron density generated through tunnel ionization of the molecules and e and m_e are the charge and mass of an electron, respectively. We note that the electron-ion recombination time is normally of the order of many nanoseconds, much longer than the femtosecond time scale of the pulse. Hence, the generated plasma could be considered as static during the interaction with the pulse. The wave enters the optical medium at ($z=0$, $t=0$). At the position z, Eq. (2.3) becomes

$$F(z,t) = \exp\left\{i\left[\omega_0 t - \frac{\omega_0 n_0}{c}z - \frac{\omega_0 \Delta n\,(t)}{c}z\right]\right\} \tag{2.5a}$$

$$= \exp\left\{i\left[\left(\omega_0 t - \frac{\omega_0 n_0}{c}z\right) + \int_0^t \frac{\partial}{\partial t}\left(-\frac{\omega_0 \Delta n\,(t)}{c}z\right)dt\right]\right\} \tag{2.5b}$$

$$\equiv \exp\left\{i\left[\left(\omega_0 t - \frac{\omega_0 n_0}{c}z\right) + \int_0^t (\Delta\omega)dt\right]\right\} \tag{2.5c}$$

where

$$\Delta\omega = \frac{\partial}{\partial t}\left(-\frac{\omega_0 \Delta n\,(t)}{c}z\right) = -\frac{\omega_0}{c}z\frac{\partial[\Delta n\,(t)]}{\partial t} \tag{2.6}$$

SPM is the modulation (variation) of the phase of the wave due to the self-generated extra phase $-\frac{\omega_0 \Delta n(t)}{c}z$ (see Eq. 2.5a). It is manifested by the frequency shift $\Delta\omega$ of Eqs. (2.5c) and (2.6). Since the front part of the pulse always sees a neutral gas, from Eqs. (2.6) and (2.5), without the plasma contribution,

$$\Delta\omega = -\frac{\omega_0 z}{c}\frac{\partial[\Delta n\,(t)]}{\partial t} = -\frac{\omega_0 z}{c}n_2\frac{\partial I(\mathit{front\ part})}{\partial t} < 0 \tag{2.7}$$

The last inequality in Eq. (2.7) arises because the front part of the pulse has a positive temporal slope whose value ranges continuously between zero and a maximum value. Hence, the front part of the pulse contributes principally to red (Stokes) shift/broadening. But the back part of the pulse should also see the neutral gas since the gas is only partially ionized. SPM in the neutral would lead to a blue shift/broadening but this blue shift is masked by the stronger blue shift/broadening due to SPM in the plasma together with the eventual SPM, due to the very steep descent of the back part of the pulse (i.e., self-steepening, Gaeta, 2000; Aközbek et al., 2001).

The contribution of the plasma term in Eq. (2.5) to frequency shift/broadening starts soon after the plasma is generated. The plasma interacts at the self-focal zone with the self-focusing slice of the pulse and with the slices coming from behind (i.e., from the back part of the pulse). Using Eqs. (2.5) and (2.6), the frequency shift/broadening due to the plasma term is

$$\Delta\omega = +\frac{2\pi z e^2}{c m_e \omega_0}\frac{\partial N_e}{\partial t} \tag{2.8}$$

The electrons are generated through tunnel ionization of the air molecules. For simplicity, we can use the experimental results to empirically state that the effective tunnel ionization rate of nitrogen (and oxygen) molecules is proportional to the intensity raised to the power of m, where m is the empirical slope of the experimental ion yield versus intensity curve in the log-log scale (Talebpour et al., 1999a). Thus, by solving the rate equation for the generation of free electrons by tunnel ionization, the electron density can be expressed as (Liu et al., 2002)

$$N_e \cong N_0 w \int_{-\infty}^{t} I^m(t)dt \tag{2.9}$$

Equation 2.9 is derived from the rate equation of tunnel ionization generating the electron density N_e in a gas, namely,

$$\frac{\partial N_e(t)}{\partial t} \propto N_0 I^m \tag{2.9a}$$

i.e., the rate of change of the electron density is proportional to N_0, the density of the neutral air and to I^m. The latter is the empirical mth power relation of the intensity I obtained from experiments. Integrating Eq. (2.9a), one obtains Eq. (2.9) where w is the proportionality constant sometimes called the "cross section" of tunnel/multiphoton ionization. Substituting Eq. (2.9) into Eq. (2.8), we obtain

$$\Delta\omega = +\frac{2\pi z e^2 N_0 w}{c m_e \omega_0} I^m(t) \tag{2.10}$$

This positive blue shift/broadening of the frequency is large partially because of the highly nonlinear dependence on the high intensity inside the self-focal region. Besides SPM, towards the end of the propagation; i.e., towards the end of the diffraction length given by ka^2 (k is the wave number and a, the beam radius at 1/e value), the back part of the intensity distribution of the pulse becomes very steep and the slope is negative. This temporal variation happens mostly in the neutral gas because it is at the end of filamentation where ionization is negligible. It would give rise to a large blue shift of the frequency since by analogy to equation 2.7

$$\Delta\omega = -\frac{\omega_0 z}{c} n_2 \frac{\partial I(\text{very steep back part with negative slope})}{\partial t} \tag{2.11}$$

This is a major source of the large broadening towards the blue side of the pump frequency. This so-called self-steepening in the case of the propagation of a powerful femtosecond pulse is the consequence of a continuous spatio-temporal self-transformation process of the pulse during propagation. Numerical simulation (see Section 4) shall give a quantitative picture of what happens (Gaeta, 2000; Aközbek et al., 2001). Here, we shall give a very qualitative, simple minded picture so as to grasp a physical feeling of self-steepening. Group velocity dispersion and the interaction with the plasma are neglected. The central slice of the pulse with the highest intensity where ionization occurs would propagate with a velocity c/n, where $n = n_0 + \Delta n_{\text{Kerr}} - \Delta n_p$. Here, n is the total index of refraction, n_0, Δn_{Kerr} and Δn_p are the indices of refraction of the neutral air, the nonlinear Kerr index and the index of the plasma, respectively. At the intensity clamping position, $\Delta n_p = \Delta n_{\text{Kerr}} = n_2 I$. This focusing central slice with intensity clamping would thus have an index of refraction $n_c = n_0$ where the subscript c indicates 'central slice'; it would propagate faster than the front part of the pulse which sees an increase of the index of refraction due to the nonlinear contribution (Kerr nonlinear index) only $n_f = n_0 + \Delta n_{\text{Kerr}}$ where the subscript f indicates front 'part'; i.e., no plasma generation yet. Now the back part of the pulse sees a weak plasma generated by the peak of the pulse. The index of refraction of this plasma zone is the combined values of the neutral and the weak plasma; i.e., $n_b = n_0 + (\Delta n_{\text{Kerr}})_b - \Delta n_p$; here, the subscript b indicates 'back part', $(\Delta n_{\text{Kerr}})_b < \Delta n_p$ because the intensity of this back slice is weaker than the clamped intensity while the plasma is left behind by the clamped intensity. Hence, $n_b < n_c < n_f$; i.e., the back part of the pulse would propagate faster than the front part of the pulse. Soon, the back part would almost catch up with the front part resulting in a steep rise in intensity at the back part. SPM is proportional to the derivative of this part of the pulse, hence a very large blue shift according to Eq. (2.11).

The propagation distance z also plays a role in both the red and blue broadening (see Eqs. 2.7 and 2.10). Thus, during experimental observations, the spectral broadening of the pulse develops progressively as the propagation distance increases. Both experiment and numerical simulation (Gaeta, 2000; Aközbek et al., 2001; Kandidov et al., 2003a) show similar broadening. A strong broadening in air towards the red up to 4 μm was recently reported (Kasparian et al., 2000b). The central part of the pattern of Fig. 2.1c, d are examples of such frequency broadening from the pump at 800 nm towards the blue side across the whole visible frequency range; hence, it appears white. This is what we call the self-transformed white light laser pulse.

2.3.8 Conical Emission

The colorful rings in Fig. 2.1a are another manifestation of self-phase modulation in the radial direction. The previous section describes self-phase modulation in the plane wave approximation; i.e., in the z-direction only. That is to say, we have considered only the wave vector k_z. However, the laser pulse front is curved at the self-focal zone. It contains a transverse part of the wave vector. In a normally spherically symmetric pulse front, the general wave vector $\vec{k}$ is given by

$$\vec{k} = k_z\hat{z} + k_r\hat{r} = k_{z0}\hat{z} + \Delta k_z\hat{z} + k_{r0}\hat{r} + \Delta k_r\hat{r} \tag{2.12}$$

where the initial wave vectors contain a subscript zero. $\hat{z}$ is a unit vector in the propagation direction; $\hat{r}$ is a unit vector transverse to $\hat{z}$. In the plane wave approximation described above, we have considered only the z-components. They are:

$$k_{z0}\hat{z} = \frac{\omega n_0}{c}\hat{z}$$

$$\Delta k_z\hat{z} = -\frac{\omega\Delta n}{c}\hat{z} = \int_0^t \frac{(\Delta\omega)}{z}dt.\hat{z} \quad \text{(from Eq. 2.5.1 and 2.5.3)}$$

$$= \begin{cases} -\dfrac{\omega_0}{c}n_2\displaystyle\int_0^t \frac{\partial I(\text{front part})}{\partial t}dt.\hat{z} < 0 \quad \text{(in neutral gas)} \\ +\dfrac{2\pi e^2}{cm_e\omega_0}\displaystyle\int_0^t \frac{\partial N_e}{\partial t}dt.\hat{z} > 0 \quad \text{(in plasma)} \end{cases} \tag{2.13}$$

$$\text{(from Eq. 2.7 and 2.8)}$$

$$= \begin{cases} -\dfrac{\omega_0}{c}n_2\displaystyle\int_0^z \frac{\partial I(\text{front part})}{\partial z}dz.\hat{z} < 0 \quad \text{(in neutral gas)} \\ +\dfrac{2\pi e^2}{cm_e\omega_0}\displaystyle\int_0^z \frac{\partial N_e}{\partial z}dz\,.\hat{z} > 0 \quad \text{(in plasma)} \end{cases} \tag{2.14}$$

$$\equiv \begin{cases} \Delta k_z\hat{Z} \quad \text{(neutral)} < 0 \\ +\Delta k_z\hat{Z} \quad \text{(plasma)} > 0 \end{cases} \tag{2.15}$$

We see that from Eqs. (2.13) to (2.14), the temporal rate of change has been transformed into a spatial rate of change by recognizing that $z = ct$. Figure 2.18a gives a schematic relationship of these waves vectors. From eq. 2.15, $\Delta k_z\hat{z}$ (plasma), being positive, is in the same direction as that of the original vector $k_{z0}\hat{z}$ while $\Delta k_z\hat{z}$ (neutral), being negative, is in the opposite direction to that of $k_{z0}\hat{z}$. Equation (2.14) (together with Eq. 2.8) shows that the spatial gradient of the electron density gives rise to a blue shift of the frequency in the z-direction (plane wave approximation). Since electrons are generated in the 3-D self-focal volume, electron density gradients show up in all directions; i.e., in both the z- and the r-directions. In the r-direction, the electron density gradient would give rise to a spatial divergence of the radiation. Thus, the wave vector, $\Delta k_r\hat{r}$ (plasma), which is in the direction of

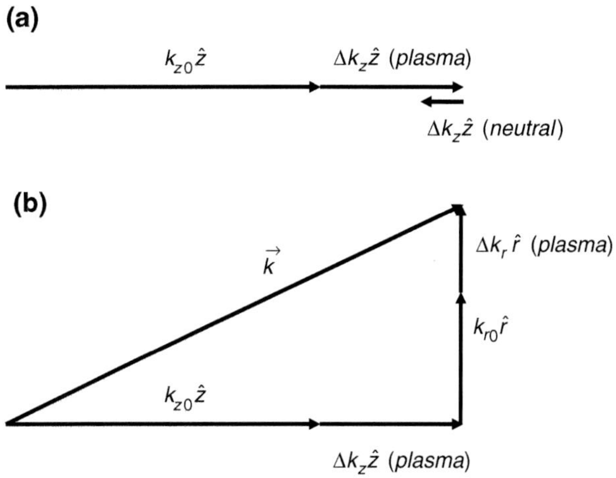

Fig. 2.18 Illustration of the various wave vectors of a spherical wave front in the self-focal region. (**a**) All the possible wave vectors in the propagation z-direction are shown. (**b**) Illustrating the transverse part of the wave vector

$k_{r0}\hat{r}$, would make this blue shifted radiation diverging into a ring as shown by the vector diagram in Fig. 2.18b. The larger the electron density gradient is, the larger will be the wave vector $\Delta k_r\hat{r} \sim \frac{\partial N_e}{\partial r}\vec{r}$ and the blue shift (Eqs. 2.8 and 2.14). The radial electron density gradient varies continuously from zero to a maximum value. The detailed analysis of conical emission is given in Kosareva et al. (1997b) and Kandidov et al. (2003a). From Fig. 2.18b, the divergence of the resultant vector $\vec{k}$ will be larger when the electron density gradient is larger. Hence, rainbow-type colored rings are generated around the central white spot; the larger the frequency shift is (i.e., the shorter the shifted wavelength is), the larger the divergence will be. The rings in Fig. 2.1d are thus explained. The frequency shift due to the neutrals would not give rise to rings because the wave vector $\Delta k_r\hat{r}$ (neutral) points in the opposite direction of $\vec{k}_{r0}$; i.e., it tends to reduce the divergence of the wave. When the above description was done numerically in 3-D calculation, the agreement between theory and experiments is excellent. Other theories such as four-wave-mixing and Cherenkov radiation have been proposed but they fail to satisfy all the experimental observations (Kandidov et al., 2003a).

Before ending this section, we comment on the repeating dark circular rings in the conical emission (Fig. 2.1d). It is due to the spectral interference (see for example: Chin et al., 1992) of the conical emissions coming from two to three on-axis filaments due to multiple refocusing (Fig. 2.5). Assume that there are two sets of conical emission from two re-focusing filaments along the same axis. Because each color of a conical emission is generated continuously along a filament, the width of each color on the observation screen is wide. Hence there will be overlap of the same color coming from the two filaments. At the overlapping positions, those colors satisfying the condition $nd = (m+1/2)\lambda$ (d = path difference between the conical light of the same color coming from two different filaments, λ = wavelength,

n = index of refraction, m = integer) would give rise to destructive interference (dark rings), while those satisfying $nd = m\lambda$ would give rise to constructive interference. If there is a bunch of filaments occurring around the propagation axis, there will be so many overlapping rings that the whole conical emission becomes a blur of continuous colored band with no dark rings. This situation can be achieved by passing a sub-terawatt pulse through a piece of BK7 glass without external focusing (Fig. 2.4). In this case, the peak power (300 GW) is much higher than the critical power for self-focusing in glass (a few megawatts) so that multiple refocusing as well as multiple filaments occur around the propagation axis.

2.3.9 Ring Structure at the Pump Wavelength

Conical emission has become an accepted name for the colored rings surrounding the filament axis as described in the previous Section 2.3.8. However, the fundamental wavelength also exhibits ring structure because of the same reason for the creation of the colored conical emission; i.e., spatial gradient of electron density around the filament. But it is not trivial to observe these rings at the pump wavelength because it is masked by the background reservoir at the same wavelength. If one uses an interference-filter to try to filter out the other wavelengths, what is left behind would be the stronger background reservoir with an irregular fluence distribution which might overwhelm the ring pattern at the pump wavelength. The author and his co-workers (Chin et al., 2002) have succeeded in doing such a measurement in air at a distance of more than 90 m away. The background reservoir, after propagating this long distance, has spread out significantly and the rings become evident. This is shown in Fig. 2.19. Note that the central white spot is the over-exposed central part of the pulse at the pump wavelength around 800 nm.

Another manifestation of the ring structure is seen by focusing the beam in air onto a piece of glass and observing the burn pattern. Damage ring structure was observed. This was confirmed by simulation (Chin et al., 2001). Figure 2.20 shows the calculated fluence distribution across the diameter of a filament in air just before the geometrical focus of a lens. Rings are evident. There is a dip at the center. These are due to the diffraction of the plasma inside the filament. Care should be taken in the interpretation of this dip. It is the fluence, not the peak intensity of the pulse. This fluence is expressed as the energy per unit surface of the cross section of the laser pulse and is the temporal integration of the intensity distribution of the pulse. We shall come back to this when we describe the theoretical simulation of the phenomena during filamentation.

2.3.10 Self-Pulse Compression

In recent years, there is a surge of interest in the compression of a laser pulse down to the single cycle level for the sake of generating single attosecond laser pulse

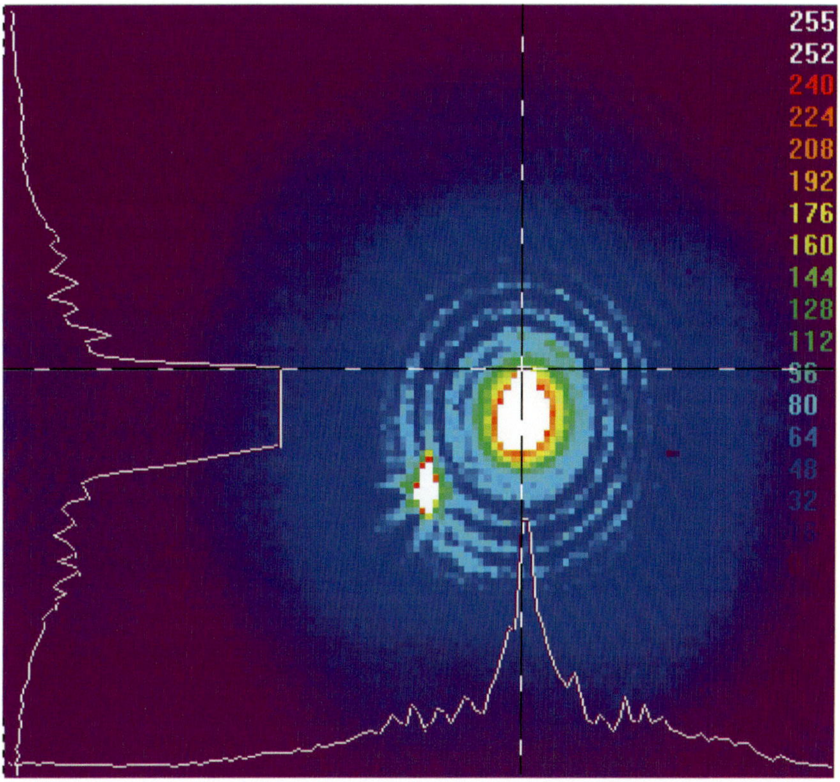

Fig. 2.19 Ring structure at the pump wavelength taken at a distance of more than 90 m away from the laser output. The *central part* is over exposed. The structure at the *lower left hand corner* is another filament that is starting to grow and interfere with the central filament. This will be dealt with in Chapter 4 on multiple filamentation

through the generation of high order harmonics in gases. It was discovered that the easiest way seems to be the technique based upon filamentation of a laser pulse (Couairon et al., 2005). This concept of self-pulse compression can be understood from the slice-by-slice self-focusing model. It is already apparent in Fig. 2.9. The front part of the pulse becomes narrower and narrower as the pulse propagates. The back part of the pulse spreads out into the broad background reservoir. At any position of the filament or after the filament, if one intercepts the pulse, one would obtain a short pulse (front part in this picture, but could also be the back part which could only be shown in numerical simulation; see later in the theory chapter) whose duration is shorter than the initial pulse duration and could become very short (one cycle). However, there is this background reservoir that has to be taken care of. Otherwise, the pulse would not be a "clean" pulse. More vivid pictures of such pulse compression will be shown in the chapter where theory and numerical simulation are discussed.

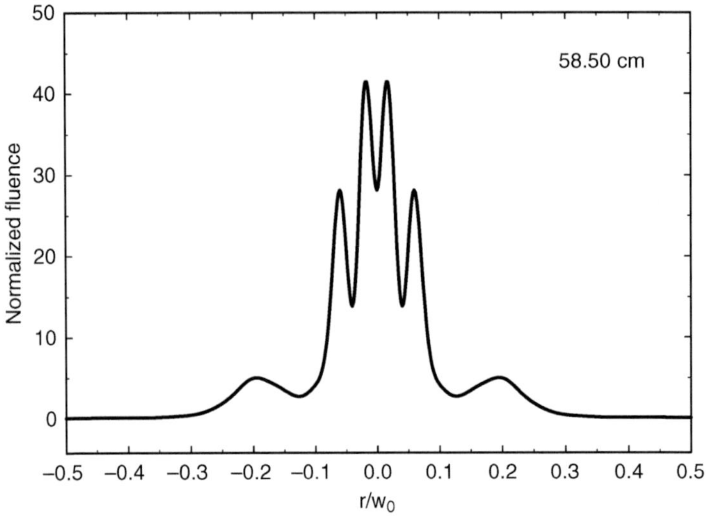

Fig. 2.20 Calculated fluence distribution across the diameter of the filament in air at $z = 58.50$ cm from a focusing lens of $f = 60$ cm (f-number = 150) and $P = 20P_c$, pulse duration = 297 fs at 1/e level of the Gaussian pulse. The laser wavelength is at 800 nm. Note the dip at the *center*. This is another manifestation of the formation of ring structure at the fundamental wavelength (courtesy of Dr. Neset Akozbek.)

2.3.11 X-wave

X-waves provide another way of presenting conical emission, supercontinuum, pulse splitting and pulse shortening in the near field (r,t) and far field (k,λ) plots of the intensity contours of the pulse at various positions of propagation during single filamentation. Here r,t,k,λ are the parameters pertaining to the pulse: radius, local time, wave vector (or cone angle) and wavelength. The intensity scale is logarithmic which amplifies the contrast in the plots thus showing some X-shaped features. Interested readers are referred to Couairon et al. (2006) for the theory and Xu et al. (2008) for an experimental manifestation of the X-wave and references therein. The physics has already been explained in the "natural" way (i.e., without using X-wave presentation) by Kandidov et al. (2003a). It will be examined in more detail in Section 3.5.

2.4 Full Evolution of a Single Filament

An experiment together with numerical simulation was done in which the full evolution of a single filament in air was measured (Chen et al., 2007). We find that the evolution of the so-called single filamentation starts with the "efficient ionization" zone. The efficient ionization zone is the usual one we measured routinely in the laboratory (Figs. 2.5 and 2.14). It is followed by the weakly ionized zone that was never observed until recently (Eisenmann et al., 2007). It was also predicted and experimentally (but indirectly) confirmed by Akturk et al. (2007). At the end of this weakly ionized zone comes the quasi-linear propagation of the pulse whose peak power is now lower than the critical power for self-focusing but whose intensity is still high enough to self-focus. However, this self-focusing is overcome by the linear diffraction of the pulse so that the divergence is very small, many times smaller than the normal low divergence of a single mode laser. After that, the pulse propagates linearly.

The experiment was done by using a collimated 10 Hz, 45 fs, 3.2 mJ, 800 nm Ti-sapphire laser beam. This beam self-focused in air at a distance. The beam diameter was measured at different positions along the propagation axis. Since the intensity was high at the self-focusing points, a wedge intercepted the beam at a near grazing angle along the filament. This ensured that the beam area on the surface of the wedge is large and hence the intensity low, thus avoiding damage of the surface. Through partial reflection, the beam pattern at the surface of the wedge was imaged onto a calibrated CCD camera with calibrated attenuation and filtering. The fluence distribution of the pattern can be obtained from the signal of each pixel from each image at each position. The beam diameter at FWHM can then be quantitatively defined. Meanwhile, the nitrogen fluorescence at different positions from inside the filament was measured from the side using appropriate UV filters and a PMT. Figure 2.21 shows the detail of the experimental setup.

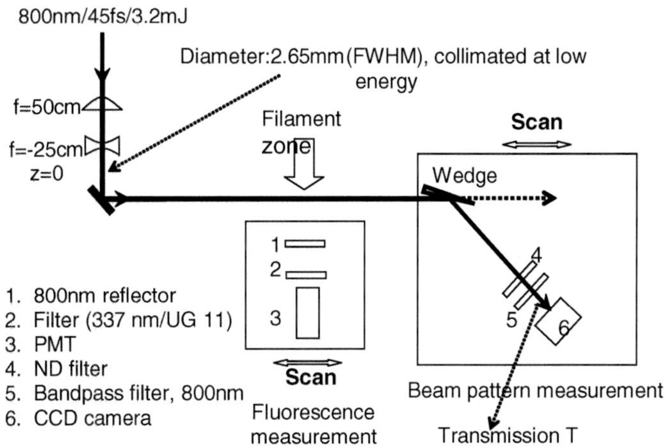

Fig. 2.21 Experimental set up to study the full evolution of single filamentation in air

Figure 2.22a–h shows the evolution of the beam's transverse pattern along the propagation axis. The central part of the beam stays round throughout the filament zone ($z = 200$–700 cm) and is always surrounded by the reservoir. This is because during filamentation, the initial pulse undergoes self-spatial filtering (Théberge et al., 2006b, 2007a; Chin et al., 2007; Liu and Chin, 2007).

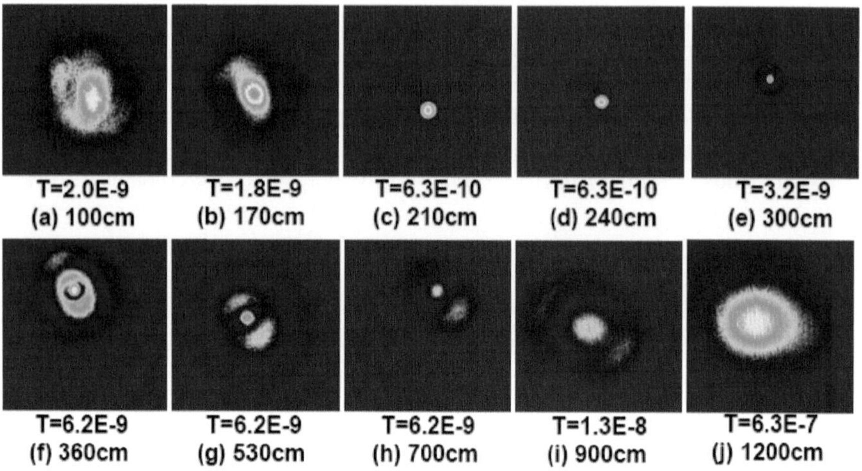

Fig. 2.22 Some typical pictures taken by CCD camera along the propagation axis. T is the transmission of the beam before reaching the CCD camera (see Fig. 2.21). The distance in cm shown below each pattern indicates the distance z from the negative lens (Fig. 2.21). The high attenuation in (c) - (e) renders the background reservoir 'invisible'

Figure 2.23a (triangular symbols) shows the fluorescence signal (log scale) as a function of z. The fluorescence signal underwent a strong decrease by two orders of magnitude from $z = 250$ to 450 cm, followed by a decrease by one order of magnitude from 450 to 690 cm. From 690 to 700 cm, the signal dropped sharply and fluctuated at noise level thereafter.

The evolution of the beam diameter (FWHM of the fluence distribution) is shown in Fig. (2.23b). A sharp decrease of the diameter until about 190 cm is followed by a slower one. The minimum of 0.18 mm is reached at around 300 cm. The beam diverges very slowly from a diameter of 0.25 to 0.37 mm at 450 cm $< z <$ 700 cm. This is the zone where the fluorescence (ionization) is weak (Fig. 2.23a).

After the filament zone, the laser pulse propagates with a divergence of 0.23 mrad (FWHM) between 900 and 1200 cm (see inset of Fig. 2.23b). This is a very small divergence since linear diffraction from the measured diameter of 370 μm at $z \cong 700$ cm would lead to a much larger divergence of $\lambda/\pi w_0 = 1.6$ mrad assuming Gaussian propagation where w_0 is the beam waistat $1/e^2$ of the peak intensity. This is because the on-axis intensity is still high after $z = 700$ cm and the pulse could still self-focus. But since the peak power is smaller than the critical power for self-focusing, there will be no collapse of the pulse. Linear diffraction together with GVD is stronger than self-focusing. The consequence is that the beam diverges slowly. The intensity in this zone soon after filamentation is estimated to be of the order of 10^{10} W/cm^2 by using the measured data described above.

Numerical simulation (Fig. 2.24) shows that at $200 < z < 300$ cm, the leading sub-pulse (peak power $\cong 3P_{cr}$) dominates (Fig. 2.24b, $z = 252$ cm). For $z > 300$ cm, the leading sub-pulse depletes; the trailing sub-pulse sustains the core energy; the

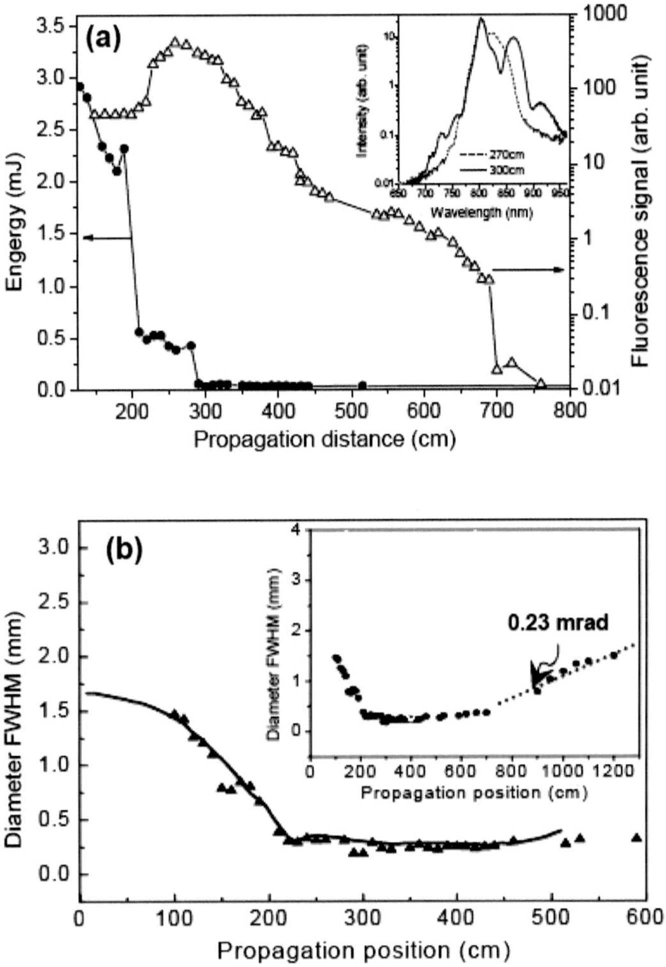

Fig. 2.23 (**a**) Measured nitrogen fluorescence signal (*open triangle*) and energy evolution of the central hot spot at the fundamental wavelength (*solid circle*) as a function of the propagation distance. *Inset*: Spectrum at position 270 and 300 cm. (**b**) Measured beam diameters (FWHM, averaged for horizontal and vertical dimension) as a function of the propagation distance (*solid triangle*) and the numerical simulations (*solid line*). *Inset*: Divergence of a nonlinear propagated beam (with filament, *solid circle*)

intermediate part diverges strongly due to the plasma left behind by the leading sub-pulse (Fig. 2.24b, $z = 308$ cm, $z = 510$ cm). As a result, the core energy decreases, while the peak power still exceeds the P_{cr}. The calculated peak intensity in the region 200–450 cm is clamped to around 5×10^{13} W/cm^2 and the peak electron density is ~0.3×10^{-4} of the atmospheric density. The filament is represented by the "efficiently ionized" zone until $z \approx 450$ cm (Fig. 2.24a). The shaded region in Fig. (2.24a), $450 < z < 620$ cm is characterized by the preservation of the peak power

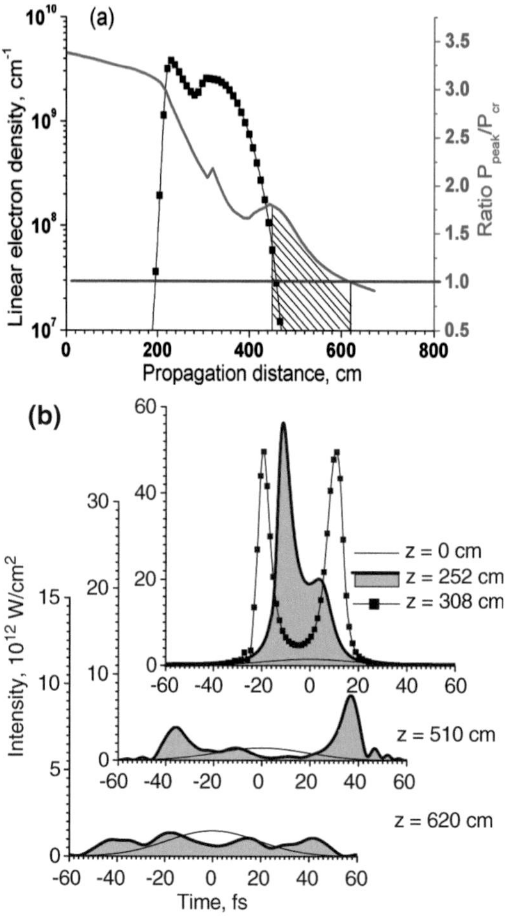

Fig. 2.24 (**a**) Linear electron density (*filled squares*) and power in the most intense slice (*solid curve*) as a function of the propagation distance. *Horizontal line* shows $P_{\text{peak}}/P_{\text{cr}} = 1$. *Dashed zone* – weakly-ionized filament. Note non-monotonic power behavior at pulse splitting position $z \approx 300\,\text{cm}$. (**b**) *Curves* with *filled pattern* are on-axis temporal intensity profiles at $z = 252\,\text{cm}$, $z = 510\,\text{cm}$, $z = 620\,\text{cm}$; *solid curve* in each plot (*not filled*) indicates initial distribution of a Gaussian pulse. *Curve*, marked by *squares*, shows pulse splitting at $z = 308\,\text{cm}$. Note change in the intensity scale for the plots at each propagation distance (courtesy of Olga Kosareva)

above P_{cr} but dominated by the trailing sub-pulse (Fig. 2.24b). The leading sub-pulse would no longer produce a plasma. In this case, the integrated electron density falls down an order of magnitude within 20 cm (440–460 cm) (Fig. 2.24a). This decrease in electron density is similar to the decrease in fluorescence in Fig. (2.23). We call the filament in this region "weakly ionized". It is important to note that the defocusing effect by plasma in this zone, the dispersion and the diffraction are still able to dynamically interplay with the self-focusing effect to sustain a self-guided column (Eisenmann et al., 2007). Later on, by $z \approx 700\,\text{cm}$, material dispersion and diffraction overcome self-focusing (Fig. 2.24a, $z = 620\,\text{cm}$) and ionization becomes negligible.

We can now understand that the so-called "nonionizing channels" at long distances might well be the local self-spatially filtered fundamental modes of various hot zones that diverge out slowly after the end of the plasma filament.

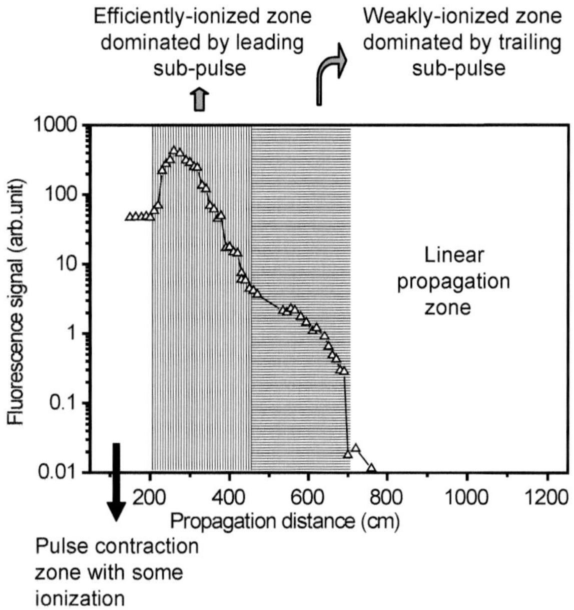

Fig. 2.25 Self-focusing and filamentation evolution in air. It starts with self-focusing. Before the collapse of the pulse, there is already some weak ionization in the pulse contraction zone. The pulse collapses into a series of self-foci (filament) with efficient ionization. This is dominated by the leading sub-pulse. The loss of energy results in the dominance of the trailing sub-pulse which continues to contribute to a much weaker ionization. At the end, the peak power of the self-filtered fundamental mode is lower than the critical power for self-focusing and it diverges/diffracts out linearly

The evolution of the so-called single filamentation can now be defined as follows (see Figs. 2.25 and 2.26): pulse contraction in the quasi-stationary regime (and self-spatial filtering) overcoming linear diffraction and material dispersion → efficiently ionized filament dominated by the leading sub-pulse → weakly ionized filament dominated by the trailing sub-pulse → weak self-focusing (no collapse) overcome by linear diffraction and material dispersion (the author calls this "quasi-linear diffraction") → linear diffraction of the self-filtered fundamental mode. There is always the background reservoir accompanying the propagation that masks the self-spatially filtered fundamental mode at the end of the propagation. That is to say, the fundamental mode and the background reservoir merge into one, and one cannot distinguish them anymore in the nonfilamentation zone.

The merging of the background reservoir and the filament core (fundamental mode) after filamentation is universal. That is to say, for a given spatial fluence distribution of a pulse, if one increases the energy inside the pulse linearly without changing any other property of the pulse, and assuming that the change does not induce multiple filamentation, the only change during propagation would be the

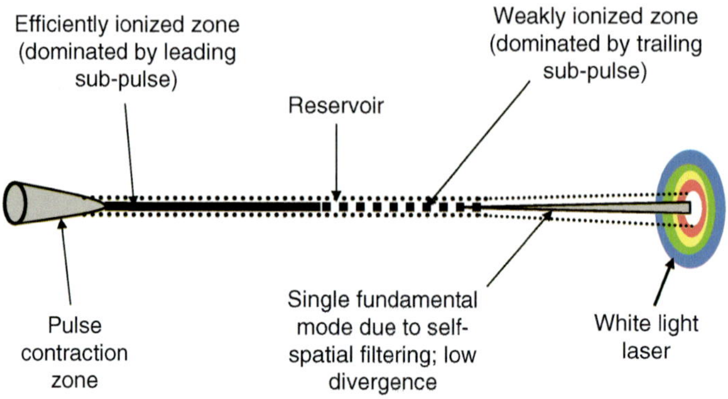

Fig. 2.26 A schematic illustration of the evolution of the filamentation of a powerful femtosecond laser pulse in air

filament length. Since the pulse would always end its filamentation when the peak power is lower than the critical power, it would always end up with the same spatial distribution (merging of the background reservoir and the fundamental mode); it would thus always diverge in the same way.

2.5 Maturity of a Filament

A filament becomes mature when it has undergone the full evolution of filamentation as shown in Fig. 2.26. That is to say, after self-focusing, the pulse evolves into the pre-filament zone, the filament zone and the post-filament zone. The back part of the pulse becomes very steep (self-steepening). It becomes a white light laser pulse. Such an evolution into a mature filament has been studied in dye solutions demonstrating clearly this evolution scheme (Schroeder and Chin, 2004).

2.6 Filamentation Without Ionization

We can now ask the question, "What really is a filament." So far, it is implicitly defined as being the self-focusing zone along the propagation axis of the laser pulse where there is ionization. There are other definitions such as using the intensity or using the burn spot on a burn paper in the laboratory as a criterion. Thus there is a debate as to what definition to use and what constitutes a filament. Méchain et al. (2004) claimed that there is filamentation without ionization at long distances (more than 1 km). They discovered that hot zones (measured by burn papers) at a long distance had an intensity less than 10^{11} W/cm^2. As it takes more than 10^{13} W/cm^2 to ionize air molecules,, it was concluded that the filaments at long distances did not generate ionization. The question is why they did not observe the higher intensity zone that ionizes air molecules at long distances (1 km away). The results in the above section (full evolution of a filament) answer this question. The so-called filamentation without ionization at long distances corresponds to the last part of the evolution; i.e., the quasi-linear diffraction of the self-spatially filtered single mode pulse right after filamentation. It has a small divergence and hence a small diameter giving rise to a respectable intensity that will give a central stronger burn spot surrounded by a weak burn pattern (reservoir) on a burn paper but this central hot spot is not intense enough to ionize air molecules anymore. We have measured the intensity to be of the order of 10^{10} W/cm^2 in this zone which is similar to the measured value of 10^{11} W/cm^2 made by Méchain et al. (2004). This slowly diverging beam propagating nonlinearly with no collapse (i.e., no ionization) would take a long distance to become linear since there is no other major mechanism to induce loss except for scattering. This explains qualitatively why Méchain et al. (2004) observed hot spots more than one km away.

2.7 What Is a Filament?

In view of the continuous dynamic processes that take place during the filamentation of a femtosecond laser pulse, a unique definition of a filament might never be found. However, because of the popularity of this new direction of research, it seems that there is a need to standardize a definition of a filament so as to avoid confusion (Chin et al, 2008b).

Intensity clamping is a unique phenomenon during filamentation. The clamped intensity is very stable. Its root-mean-square fluctuation is more than 10 times less than that of the input pulse before filamentation (Théberge et al., 2006b). The author proposes using intensity clamping as a guideline for the definition of a filament. Because intensity clamping is accompanied by plasma generation, it is convenient to define a filament as being the zone where there is plasma generation. Thus, in the context of the current study, the efficiently and weakly ionized zones together constitute the filament.

Because intensity clamping is a unique phenomenon, the author proposes the following definition. A filament is the propagation zone where there is intensity clamping. Since intensity clamping is the consequence of the generation of plasma that balances the self-focusing act of the pulse, this definition would mean that a filament is the propagation zone where plasma is generated. The zones before and after the plasma zone where the intensity is still relatively high but not sufficient to generate a plasma could be conveniently called pre- and post-filament zones.

Before ending this section, it should be emphasized that self-focusing alone is not always sufficient to produce a filament. To obtain self-focusing, it is sufficient to have an inhomogeneityof intensity across the wave front. But if the peak power is lower than the critical power for self-focusing, or if the divergence of the beam is stronger than the focusing power of self-focusing (Liu et al., 2006a), linear diffraction, GVD and divergence would overcome the self-focusing and there will be no collapse that leads to filamentation.

Chapter 3
Theory of Single Filamentation

3.1 Introduction

The propagation of a femtosecond optical pulse is described by Maxwell's equations. A number of linear and nonlinear effects such as self-focusing, dispersion, self-phase modulation, and ionization have to be taken into account. The overall dynamics of such pulses can be complicated where both transverse and temporal effects play equally important roles (Chin et al., 2005; Couairon and Mysyrowicz, 2007; Bergé et al., 2007 and references therein). Unfortunately, so far there is no analytical solution to the problem. Finding direct numerical solutions with a computer requires enormous computational efforts and in many cases does not provide an insight to the basic physical understanding of the various linear and nonlinear effects involved. Therefore, an approximate wave equation is used instead with a more reasonable mathematical solution. The interaction of intense optical pulses in a bulk medium is highly nonlinear and the material response must therefore couple self-consistently with the wave equation. The following is adapted from Chin et al. (2005).

3.2 Filamentation in Air

We consider propagation in air as a concrete example. From Maxwell's equations it is possible to obtain a second order scalar wave equation for the electric field. This scalar equation is obtained by assuming that we have a linearly polarized electric field, E, propagating in the medium which is assumed isotropic. The isotropic nature of the medium is assumed unchanged during filamentation. The vector nature of the field is thus suppressed in writing the expression with the understanding that it is linearly polarized. The derivation of this equation can be found in any book on electromagnetic theory. It reads (in Gaussian units)

$$\frac{\partial^2 E}{\partial z^2} + \frac{\partial^2 E}{\partial x^2} + \frac{\partial^2 E}{\partial y^2} - \frac{4\pi}{c^2}\frac{\partial^2 P}{\partial t^2} - \frac{4\pi}{c^2}\frac{\partial J}{\partial t} = 0 \tag{3.1}$$

S.L. Chin, *Femtosecond Laser Filamentation*, Springer Series on Atomic, Optical, and Plasma Physics 55, DOI 10.1007/978-1-4419-0688-5_3,

where P is the polarization response of the medium and includes both linear and nonlinear responses of the medium. The current density J comes from free electrons created by field/tunnel ionization. It can be written as

$$J = -eN_e v_e. \tag{3.2}$$

where the electron velocity ν_e is derived from

$$\frac{\partial v_e}{\partial t} = -\frac{eE}{m_e} \tag{3.3}$$

Let us take the time derivative of Eq. (3.2) and assuming that the initial electron velocity is zero when it is created, we obtain

$$\frac{\partial J}{\partial t} = \frac{e^2 N_e}{m_e} E \tag{3.4}$$

Inserting Eq. (3.4) into Eq. (3.1) we obtain

$$\frac{\partial^2 E}{\partial z^2} + \frac{\partial^2 E}{\partial x^2} + \frac{\partial^2 E}{\partial y^2} - \frac{4\pi}{c^2}\frac{\partial^2 P}{\partial t^2} - \frac{4\pi e^2 N_e}{c^2 m_e} E = 0 \tag{3.5}$$

First we consider laser pulse propagation in the framework of the slowly varying envelope approximation (SVEA) and will discuss the effects of higher-order correction terms such as self-steepening later. We assume that the medium polarization is given by $P = \chi^{(1)}E + \chi^{(3)}E^3$, where $\chi^{(1)}$ and $\chi^{(3)}$ describe the linear and nonlinear susceptibility coefficients, respectively. The dielectric function is given as $\varepsilon = 1 + 4\pi\chi^{(1)}$. The electric field is assumed to have a rapidly oscillating part $e^{-i\omega t + ikz}$ that is modulated by an envelope given as $E(x,y,z,t) = \varepsilon(x,y,z,t)e^{-i\omega t+ikz} + c.c.$, where $\varepsilon(x,y,z,t)$ is assumed to be a slowly varying envelope function such that it varies slowly in time and space on the scales of ω^{-1} and k^{-1}. Inserting P and E into Eq. (3.5) and applying the slowly varying envelope approximation one obtains

$$i\frac{\partial \varepsilon}{\partial z} + \frac{1}{2k}\nabla_\perp^2 \varepsilon + n_2 k_0 |\varepsilon|^2 \varepsilon - \frac{2\pi e^2 N_e}{k m_e c^2}\varepsilon + i\Gamma\varepsilon = 0 \tag{3.6}$$

Here, $\nabla_\perp^2 = \frac{\partial^2}{\partial x^2} + \frac{\partial^2}{\partial y^2}$ is the transverse Laplacian operator. In addition, for simplicity, effects arising from the group velocity dispersion are neglected and can be added easily as is done later. Equation (3.6) is the so-called nonlinear Schrödinger equation. The intensity of the pulse is defined as $I = |\varepsilon|^2$ given in units of W/cm^2, and $n_2 = 12\pi^2\chi^{(3)}/n_0^2 c$ is the nonlinear coefficient given in units of cm^2/W and Γ accounts for ionization losses. The generation and evolution of electron density $N_e(z,x,y,t)$ for single ionized molecules is given by

$$\frac{\partial N_e}{\partial \tau} = (N_0 - N_e)R(|\varepsilon|^2) \tag{3.7}$$

where R is the multiphoton/tunnel ionization rate (in units of s^{-1}) for air molecules (oxygen and nitrogen) and N_0 is the number density of neutral molecules in units of molecules/cm^3. Effects such as electron recombination and cascade ionization, which do not play an important role for pulses shorter than 1 ps (Section 1.4), are neglected.

Equation (3.6) when coupled to Eq. (3.7) includes the most basic linear and nonlinear terms for describing self-focusing, diffraction, plasma generation and defocusing, leading to the concept of laser pulse filamentation. It is referred to as the nonlinear Schrödinger equation and has been studied widely in nonlinear optics. As we will show later, the slowly varying envelope approximation eventually breaks down but nevertheless provides an important physical understanding of these phenomena.

3.3 Numerical Solution of Filamentation in Air

We shall analyze the numerical solution and compare with known experimental results. We consider the propagation in air of an initially collimated Gaussian beam of the form $\varepsilon(r,\tau) = A_0 e^{-(r^2/w_0^2+\tau^2/\tau_0^2)}$, where w_0 and τ_0 are the initial beam radius and pulse width respectively, measured at $1/e^2$ of the intensity (Fig. 3.1a). Equations (3.6) and (3.7) are integrated with $w_0 = 0.025$ cm, $\tau_0 = 85$ fs and $P_0 = 6P_{cr}$, where $P_{cr} = 3$ GW is the critical power for self-focusing in air at 800 nm. In addition, cylindrical symmetry of the transverse profile is assumed which is valid when describing the evolution of a single filament. In Fig. 3.1, the time scale is the local time of the pulse. Negative time means the front part of the pulse. We can also imagine that the pulse is propagating in space in the z-direction ($z = ct$ where c is the speed of light and t is time). In this case, z is toward the direction of negative time. (It should be noted that the critical power for self-focusing is pulse-duration dependent in air because of the molecular delayed response (Liu and Chin, 2005). For about 100 fs or longer pulses, P_{cr} is around 4 GW. It is higher for shorter pulses. It could be as high as 10 GW in the case of 45 fs pulses. In this chapter, the "old" value, 3 GW, is kept. It would not change the physical understanding of the physics).

Electron generation through multiphoton/tunnel ionization of nitrogen and oxygen molecules (assuming 80% nitrogen and 20% oxygen) is taken into account via the rates R in Eq. (3.7). The rate R is fitted by the empirical formula $R = \sigma^{(n)} I^n$, where $\sigma^{(n)}$ and n are fitting parameters obtained through experimental measurement of the ionization of oxygen and nitrogen molecules separately using 200 fs Ti-sapphire laser pulses (Talebpour et al., 1999a). The experimental results are plotted on log-log scales of the number of ions generated through tunnel ionization vs. laser intensity. For the relevant intensity range up to about 10^{14} W/cm^2, the plot is a straight line whose slope indicates the exponent n (Chin, 2004). Unlike the result of multiphoton perturbation theory, this n is, in general, not an integer number because the ionization process is in the tunneling regime (Chin, 2004; Chin et al., 1985).

A more theoretical approach is to use the "Intense-Field Many-Body S-Matrix Theory" (IMST) to calculate the ionization rate (Becker and Faisal, 2005). This theory was found to agree very well with the experimental data obtained by Talebpour et al. (1999a). But since during filamentation, the pulse gets shorter and shorter, the ionization physics would evolve into the few-cycle regime where the experimental results might not be the same. However, the ionization rates obtained using the IMST are still valid as long as the pulse width at half the maximum intensity (FWHM) is at least three field cycles long (Becker et al., 2001b). (This is the reason why in the simulation described above, the experimental data are used for the sake of simplicity).

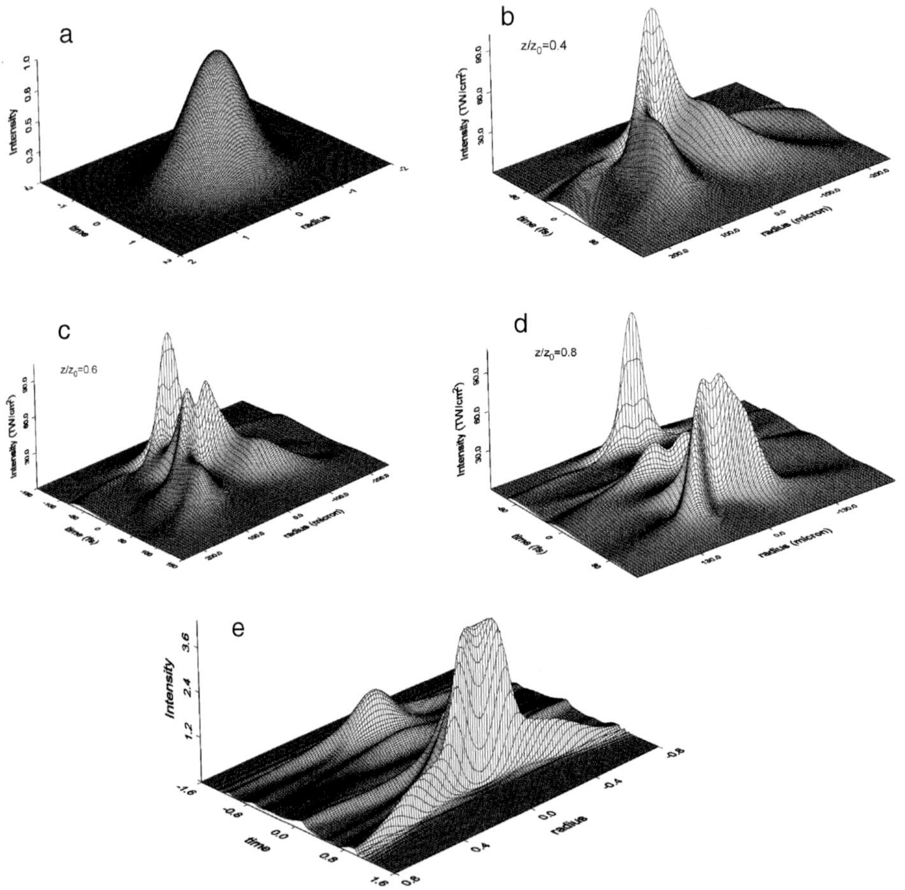

Fig. 3.1 Evolution of a femtosecond laser pulse propagating in air. Shown is the spatio-temporal intensity distribution of (**a**) an initially Gaussian pulse propagating in ionizing air at $z = 0$, (**b**) $z/z_0 = 0.4$, (**c**) $z/z_0 = 0.6$, (**d**) $z/z_0 = 0.8$ and (**e**) $z/z_0 = 1$. The intensity is normalized to the peak input intensity of (**a**) and the radius and time coordinates are scaled to the initial beam radius $w_0 = 0.025$ cm, pulse width $\tau_0 = 85$ fs and $P_0 = 6P_{cr}$, where $P_{cr} = 3$GW (courtesy of Dr. Neset Aközbek)

Figure 3.1a–e gives the results of simulation in which the spatio-temporal intensity distribution is plotted at various propagation distances z normalized to the diffraction length $z_0 = kw_0^2/2$ of the collimated input beam (Aközbek et al., 2001). As the pulse self-focuses, the peak intensity increases very rapidly until there is enough plasma to stop the focusing process. The peak (strongest part) of the pulse will come to a focus first and will be stopped by the defocusing of the self-induced plasma as shown in Fig. 3.1b (intensity clamping). Note again that the front part of the pulse is in the negative local time region. This is followed and repeated by other parts at the front part of the pulse; i.e., "slice-by-slice self-focusing" as discussed in Chapter 2. The plasma generation is a cumulative process and each slice experiences a different magnitude of plasma defocusing. Thus some of the earlier slices will reach a higher peak intensity before being defocused. These time-dependent focusing and defocusing processes lead to the temporal reshaping of the pulse. As seen in Fig. 3.1b, there is a sharp leading edge with a smoother and very broad back component. This latter spatially broad low energy distribution of the field constitutes partially the background reservoir. It will exist persistently in all the following results so long as there is self-focusing. The front part of the pulse is now shorter than the original pulse length because the back part is diffracted by the plasma (Fig. 3.1b). This is how self-pulse compression comes by. With further propagation a second pulse appears at the back of the leading pulse which becomes even shorter, as seen in Fig. 3.1c and d. This is what has been called pulse splitting. We should call this refocusing. That is to say, the back part of the pulse, after being diffracted by the plasma left behind by the front part of the pulse, contains sufficient energy and power so that self-focusing restarts after some distance of propagation. The peak at the back part of the pulse keeps increasing as the pulse keeps propagating as shown in Fig. 3.1d.

In Fig. 3.2a, the normalized filament energy (defined as the energy contained in the filament core within a diameter of 150-μm to the total pulse energy) is plotted as a function of the normalized propagation distance, z/z_0 (Aközbek et al., 2001). Initially, due to self-focusing, more energy of the pulse is channeled into the core region until there is enough plasma generated to stop the self-focusing process, and the beam starts to defocus. However, the defocusing is stopped and the pulse refocuses again, which can be seen as the second peak in the filament energy. This process can repeat itself many times which is apparent from Fig. 3.2a as a weak third peak in the filament energy. This multiple refocusing phenomenon was observed experimentally (Fig. 2.5 and see also Chin et al., 2005). The simulation and the experiment are in good qualitative agreement with each other.

Figure 3.2b shows the generated linear electron density along the propagation direction. It clearly agrees with the refocusing discussed in the filament energy description in Fig. 3.2a. Whenever the pulse refocuses, more electrons are being generated which are seen as the peaks in Fig. 3.2b and their location agrees well with the peaks in the filament energy depicted in Fig. 3.2a. Alternatively, one can examine each temporal slice of the temporal intensity profile as a function of propagation distance. Figure 3.2c shows the on-axis intensity $I(r = 0, \tau = 0)$ of the temporal slice which has the highest initial peak power. It will come to a focus

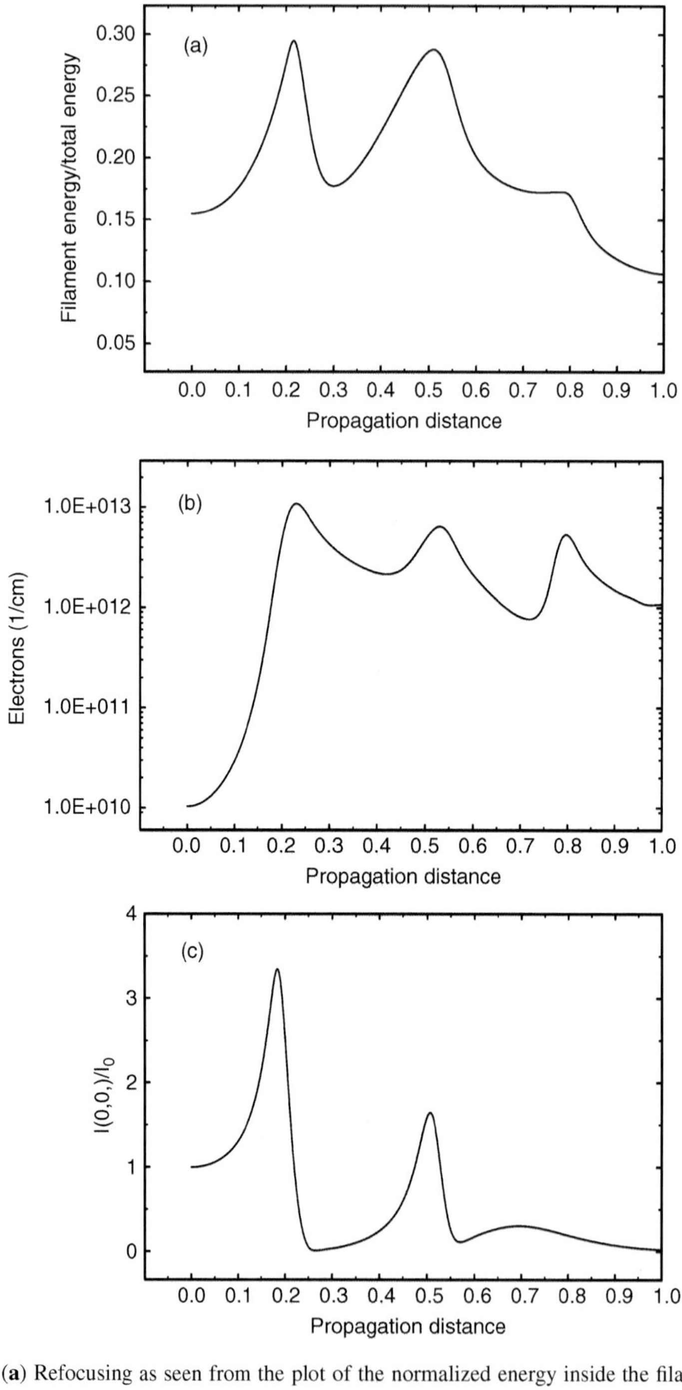

Fig. 3.2 (**a**) Refocusing as seen from the plot of the normalized energy inside the filament core versus the normalized propagation distance z/z_0. The normalized energy is defined as the ratio of the energy contained inside a diameter of 150 μm centered around the filament axis to the total pulse energy. (**b**) Re-focusing as seen from the electron density plot. (**c**) Re-focusing as seen from the intensity plot (courtesy of Dr. Neset Akozbek)

first and the peak intensity increases until plasma defocusing stops the self-focusing process and it starts to defocus; but it only defocuses until self-focusing takes over again. This process can take place many times resulting in multiple self-focusing collapses. The refocusing of the pulse channels energy back into the core of the beam and thus represents the process in which energy is exchanged between the core and the outer part (background reservoir) of the beam. This is one of the important physical mechanisms of the long-range propagation and filament formation in air.

At the end of the propagation when $z/z_0 = 1$, the front part of the pulse becomes less important than the back part of the pulse. The latter becomes very steep at the back side with a negative slope; i.e., self-steepening. This is seen in Fig. 3.1e. This steepness naturally gives rise to a very broad spectrum toward the blue side of the main frequency of the initial pulse. That is, it will result in a huge spectral broadening of the pulse; i.e., a supercontinuum or white light laser (see Chapter 2.). This is when the filament becomes mature (Section 2.5).

It is worth emphasizing that the popularly called supercontinuum is not a separate pulse generated by filamentation but is simply the same pulse which self-transforms into a white light laser pulse with a very steep back part.

Many other methods have been used to derive the nonlinear Schroedinger equation in one form or another (see the reviews in Chin et al., 2005; Couairon and Mysyrowicz, 2007; Bergé et al., 2007; Kasparian and Wolf, 2008 and the references therein). All of them give similar results as explained above.

3.4 Filamentation in Condensed Matter

In principle, the physics of filamentation in condensed matter is similar to that in air or other gases. One major difference is the generation of free electrons at the self-focus. Instead of pure tunnel ionization of gas molecules, in condensed matter, the generation of free electrons starts from the excitation of electrons from the valence to the conduction bands (Brodeur and Chin, 1998) followed by inverse Bremsstrahlung and a few cycles of collisional ionization because the density of condensed matter is high (see Chapter 1). Equation (3.7) will have to be modified. For more details, see Chin et al. (2005) and Kandidov et al. (2003a).

3.5 x-Wave and Conical Emission[1]

Recently, there have been many papers treating filamentation as x-waves generation (see for example, Di Trapani et al., 2003a; Kolesik et al., 2004). Strictly speaking, x-wave is a class of mathematical solution of the wave equation under certain constraints and is related to the solution of solitary/localized wave packets (Claudio Conti, 2005). It could be linear or nonlinear. In the nonlinear optical case, it is related to the so-called light bullet first coined by Silberberg (1990). The reason

[1]This section was written with the kind collaboration of Dr. Olga Kosareva of Moscow State University.

why it is called x-wave is probably because when the solution is expressed as a kind of contours of "intensity" distribution in a 2D plot with two parameters (normally spatial-temporal or spatial-spectral), the resulting picture (2D distribution) looks like an x-letter in the ideal case. Such x-shapes indicate the conical structures of the wave-object with cylindrical symmetry in the propagation direction.

In the case of filamentation, Kosareva and Kandidov of the Moscow State University (see Kosareva et al., 1997a, b) are among the firsts, if not the first, who have expressed the solutions of filamentation in the form of x-shaped structures (see also Golubtsov et al., 2001; Kandidov et al., 2003a, 2004). However they did not use the word x-wave because the focus of their work was not on localized wave packet. Rather, they used the 3D representation of the pulse's spatial-temporal as well as spatial-spectral intensity distribution to illustrate the physics of filamentation, white light laser (supercontinuum) and conical emission.

Let us look at their computer solution of the wave equation. Figure (3.3, top) shows, in a 3D semi-logarithmic plot, the spatial-spectral intensity distributions

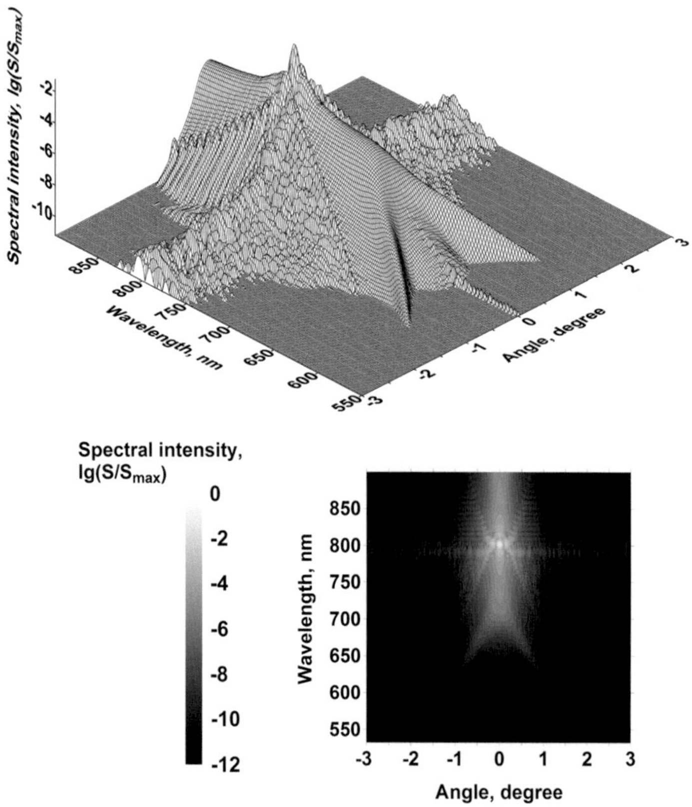

Fig. 3.3 Spatial and spectral intensity distribution of a 230 fs(FWHM)/800 nm pulse of radius $a = 167\,\mu\text{m}$ after propagating a distance of $z = 52$ m in the atmosphere. *Top*: 3D representation; *bottom*: 2D representation. For detail of the propagation physics, see Golubtsov et al. (2001, Fig. 7). This is an original unpublished plot generously given to the author by Olga Kosareva, Moscow State University

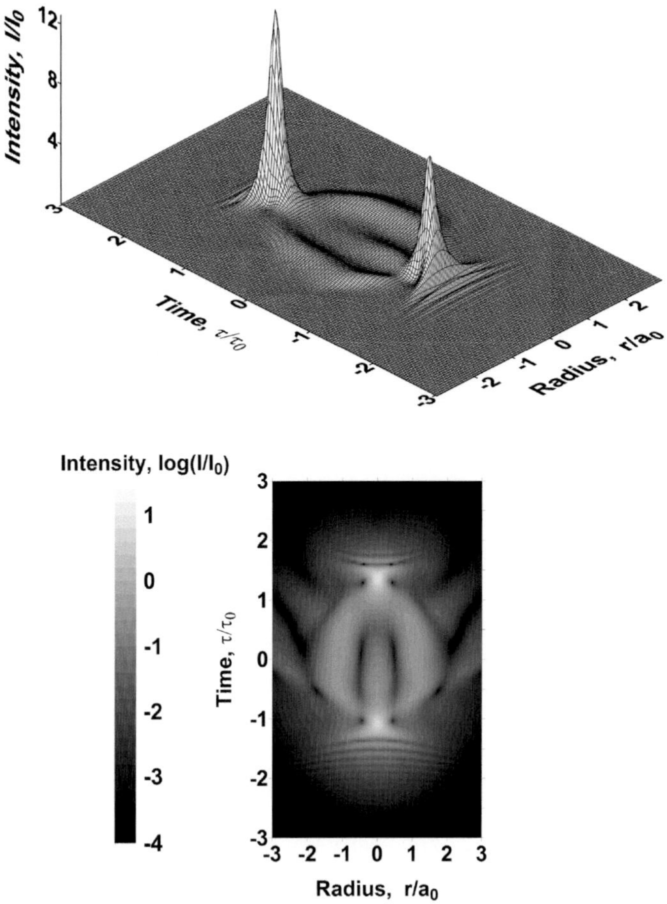

Fig. 3.4 Spatial and temporal intensity distribution of the same pulse at propagation distance of $z = 52$ m in air. *Top*: 3D representation (in linear scale); *bottom*: 2D representation (in semi-log scale). Note the x-nature of the plot. For detail of the propagation physics, see Golubtsov et al. (2001, Fig. 7) (courtesy of Olga Kosareva, Moscow State University)

$S(\theta, \lambda)$ of a fs laser pulse after propagating in the atmosphere for 52 m in the z-direction ($z = 52$ m). The normalized intensity $S(\theta, \lambda)$ in log-scale is expressed as function of the divergence angle (transverse spatial coordinate with cylindrical symmetry around angle zero) and the wavelength (spectral coordinate). This (as well as Fig. 3.4 to be discussed later) is an extension of the 3D-plots in Golubtsov et al. (2001, Fig. 7) and was done by Dr. Olga Kosareva of the Moscow State University. There is an extension of the intensity scale down to 10^{-11} showing a long tail at the short wavelength side in Fig. 3.3. In the original paper (Golubtsov et al., 2001), this scale was only down to 10^{-10}.

This 3D x-shape (or rather v-shape) structure characterizes conical emission. Conical emission of different colors (see Chapter 2.) is obtained in Fig. 3.3 (top).

Conical emission in the 3D plot (top) can be visualized by recognizing that the 3D object has a radial symmetry. Making a revolution of the plot around the central axis ($\theta = 0$) would give rise to some ring structures in the 3D plot of Fig. 3.3 (top). Ring structure corresponding to the major v-shape tail shows up at the shorter wavelength side of the fundamental wavelength (800 nm). This corresponds to conical emission toward the blue side of the fundamental as discussed in Chapter 2.. These rings at the blue side of the fundamental (see Fig. 2.1) come predominantly from the interaction and diffraction by the plasma at the front part of the pulse (Kandidov et al., 2003a, Fig. 4). Toward the red side of the fundamental, there are weaker v-shape local maxima indicating propagation of the pulse in the Kerr medium with material dispersion giving rise to rings (conical emission) (Luther et al., 1994). There should also be rings symmetrically located at the blue side of the fundamental due to the same effect if no plasma were generated. However, because of plasma generation, rings on the blue side are due to the combined effect of the dispersion and the plasma; therefore we do not have complete symmetry between the blue and the red sides of the spectrum.

Now, when the 3D plot is projected into a 2D contour plot, a complex x-shape structure, stronger at the blue side and weaker at the red side of the fundamental is revealed (Fig. 3.3, bottom). This x-shape object in filamentation is called x-wave by some (see for example, Di Trapani et al., 2003a) in an attempt to relate it to the universal dream of non-spreading wave packet. One should note that the x-tails are very weak because the plots are in log-scale. We also note that the x-wave in Fig. (3.3, bottom) looks almost the same as the x-wave obtained in water by Kolesik et al. (2004, Fig. 4). The difference between Fig. (3.3, bottom) and Fig. 4 from Kolesik et al. (2004) is mainly due to lower material dispersion in air. In particular, in air, we may not see the secondary system of "horns" at the red side of the fundamental, which, although weak, are clearly pronounced in Kolesik et al. (2004; see below).

The two secondary, weaker "horns" at the blue side of the fundamental appearing in the 2D representation or the secondary system of rings in 3D representation are associated with the trailing sub-pulse at this position of propagation of $z =$ 52 m. Figure 3.4 (top) shows the intensity distribution in space and time in a 3D plot. The peak intensity of the trailing sub-pulse is higher than that of the leading sub-pulse at this propagation distance. Thus, the trailing sub-pulse creates its own system of (weak) conical rings. In principle, the trailing sub-pulse should create rings toward both the blue and red sides of the fundamental. However, the trailing sub-pulse "front" is not so steep as the real front of the leading sub-pulse and the dispersion in air is low. Therefore, secondary "horns" at the red side of the fundamental in air are absent. If the dispersion in air were stronger, one would have exactly the same picture as Kolesik et al. (2004) who had chosen water for their simulations and x-wave generation. Note that each sub-pulse whose intensity growth is stopped by either material dispersion or the plasma forms its own x-pulse in the spatial-spectral domain. However, the ring systems from different sub-pulses may not coincide exactly. This is because the sub-pulses are produced at different propagation distances and have slightly different divergence.

The projection of the 3D representation onto a 2D representation shows again a quasi x-shape object,; i.e., again an x-wave, now in the spatial-temporal domain (Fig. 3.4, bottom). In fact, even before pulse splitting, the 2D representation of the filamenting pulse already shows the x-wave structure in the simulation by Kosareva et al. (1997a, Fig. 2).

Along the axial direction toward the shorter wavelength side in the 3D plot of Fig. (3.3, top), there is a long on-axis tail mainly due to self-steepening (Kandidov et al., 2003a, Fig. 4; Aközbek et al., 2001). Thus, the combined on-axis spectrum would be very broad and would appear as a "white" spot on the transverse pattern (Fig. 2.1) after filamentation and is the basis for what we call by "white light laser" or supercontinuum. At the fundamental wavelength in the 3D plot (Fig. 3.3, top), there is an extended ridge with many local maxima extending toward large diverging angles. These local maxima would represent the weak ring structure of the fundamental after propagation (see Fig. 2.19).

In addition, if the equi-intensity contours in the 2D plots are changed into color coding, thanks to modern computational advancement, the 2D plots become beautifully attractive turning into a piece of art (Di Trapani et al., 2003b). But they are still the same spreading pulse during filamentation. At this stage, no more new physics could be learned from the x-wave representation from the point of view of filamentation, since we have already discussed all the features of such propagations including conical emission.

Again, interested readers are referred to Couairon et al. (2006) for a review of the theory of x-waves in filamentation and Xu et al. (2008) for an experimental manifestation of the x-wave and references therein. More detailed physics has already been explained in the "natural" way (i.e., without using x-wave presentation) by Kosareva et al. (1997a, b), Kandidov et al. (2004), Golubtsov et al. (2001), Kandidov et al. (2003a), Chin et al. (2005), Couairon and Mysyrowicz (2007), and Bergé et al. (2007).

Chapter 4
Multiple Filamentation

4.1 Introduction

So far, we have discussed single filamentation with re-focusing if the peak power is high. That is to say, the beam quality of the laser pulse is assumed to be very smooth. The transverse fluence distribution of the pulse is similar to, if not equal to a Gaussian distribution in the experiment and is assumed Gaussian in the simulation. Only one filament occurs if the peak power is not more than 2–3 critical power in an experiment whereas in the simulation, a single filament always occurs. However, once the peak power is higher than about 3 critical power in an experiment, unavoidable spatial irregularity across the wave front would be induced due to the imperfection of the beam quality or due to propagation through a non-homogeneous medium in the real environment. This would result in self-focusing of the local higher intensity zone resulting in the formation of more than one filament; i.e., multiple filamentation takes place.

4.2 Multiple Filamentation: Experimental Observation

Multiple filamentation (Bespalov and Talanov, 1966) will occur during the propagation of a short laser pulse in a transparent medium whenever the intensity distribution of the wave front is not uniform. This is because a higher intensity hot/warm zone will induce a local higher nonlinear index of refraction due to the higher local intensity. This results in local self-focusing as if this were undergoing single filamentation. However, if there are two or more hot/warm zones in the pulse-front, each of them will tend to self-focus into a filament. Since each filamentation gets its energy supply from the background reservoir through the slice-by-slice self-focusing mechanism, these hot/warm zones are bound to compete for the energy from the reservoir. Moreover, each self-focus of a filament from one hot zone will radiate conical waves of different frequencies at different positions. These conical waves will interfere with those from the other filaments and the plane wave from the reservoir.

S.L. Chin, *Femtosecond Laser Filamentation*, Springer Series on Atomic, Optical, and Plasma Physics 55, DOI 10.1007/978-1-4419-0688-5_4,

An experiment is designed to observe multiple filamentation during long distance propagation in air. The experimental setup is shown in Fig. 4.1. The Ti-sapphire laser beam coming out of the compressor passes through a 10 m stainless steel vacuum pipeline and a CaF_2 window (1 cm thick) before entering the corridor next to the laboratory of the author. After a short distance (about 1 m) of deviation by mirrors, the beam propagates along the corridor. The single shot images of the patterns on the screen are imaged onto both the digital camera (colored pictures) and the CCD camera. These cameras are fixed relative to the white paper screen on a movable table. Appropriate filters are used in front of the cameras. Figure 4.2 shows the resultant colored patterns at different distances of propagation. At the beginning of propagation, the pattern is weak. The pale purple pattern is due to the fluorescence from the white paper showing the irregularity of intensity across the pattern. The fluorescence is the result of a few photon absorption by the paper. Each picture is from a different shot. Each of the hot zones undergoes filamentation independently. Some of them become "mature" earlier than the others. The higher the local peak power is at a local hot zone, the earlier filamentation starts and the earlier it degenerates into a mature filament. Each white light laser source diverges and undergoes spectral interference (i.e., interference between the same wavelengths) with the other more

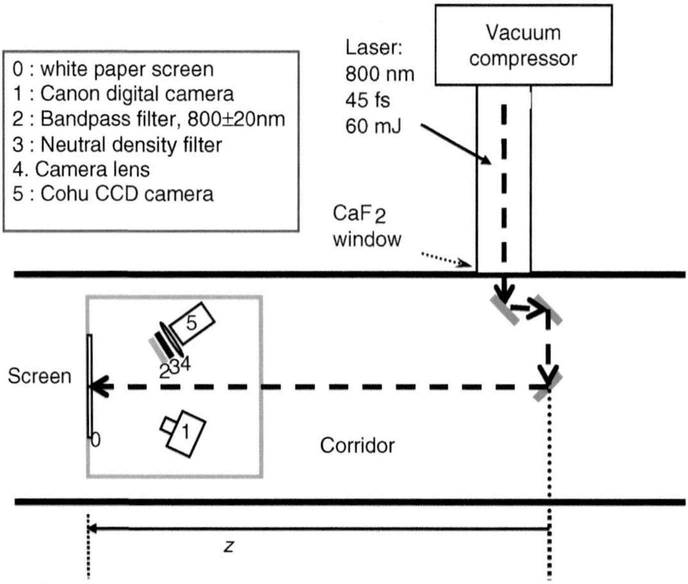

Fig. 4.1 Experimental set up to measure the patterns of multiple filamentation in air. The laser beam coming out of the compressor passes through a 10 m stainless steel vacuum pipeline and a CaF2 window (1 cm thick) before entering the corridor next to the laboratory of the author. After a short distance (about 1 m) of deviation by mirrors, the beam propagates along the corridor. The single shot images of the patterns on the screen are imaged onto both the digital camera (colored pictures) and the CCD camera. These cameras are fixed relative to the white paper screen on a movable table

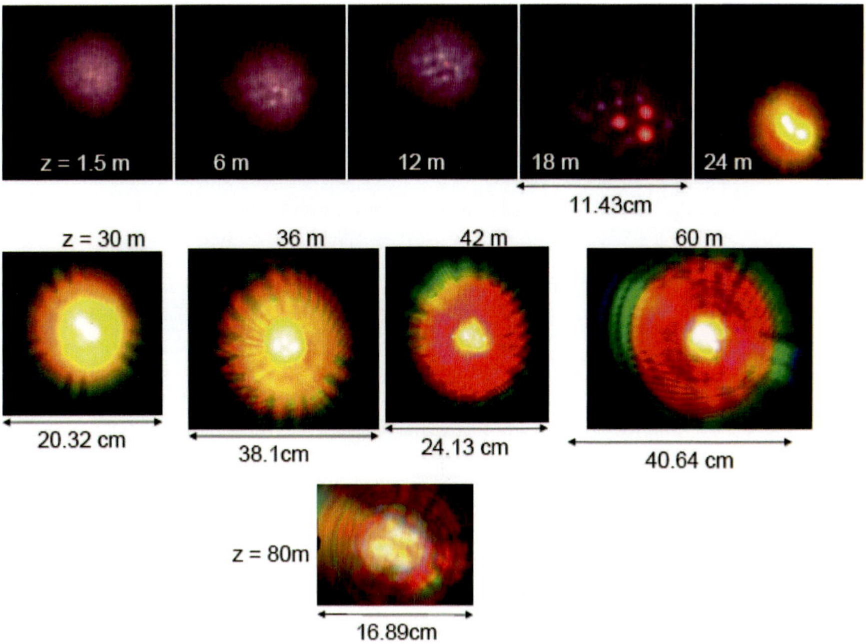

Fig. 4.2 Colored single-shot pictures showing the evolution of the patterns at different distances z from the last mirror after the vacuum pipeline. The sizes of the pictures are different. The *top row* has the same dimension. The dimensions are indicated by the *double arrows*. The beam pattern at $z = 1.5$ m already shows an irregular intensity distribution. Each picture is from a different shot. Each of the hot zones undergoes filamentation independently. Some of them become 'mature' earlier than the others. The higher the local peak power is at a local hot zone, the earlier filamentation starts and the earlier it degenerates into a mature filament. Each *white* light laser source diverges out and undergoes spectral interference with the other more or less mature sources. Constructive and destructive interferences give rise to the star like structures similar to Fig. 2.19. The picture at $z = 80$ m shows the central part of the whole pattern which becomes too large to be imaged properly.

or less mature sources. Constructive and destructive interferences give rise to the star-like structures similar to the one shown in Fig. 2.19 and will be discussed in more detail later in this chapter. The picture at $z = 80$ m shows the central part of the whole pattern which becomes too large for the camera to capture.

When we take the image of the pulse using the 800 nm band pass filter at various propagation distances, we observe those shown in Fig. 4.3. The reservoir is essentially the whole (weak) background pattern while the hot spots are evident. What is interesting is that the pattern does not diverge too much. Most of the energy of the pulse is still contained inside the 800 nm part of the pulse. However, when the pulse length was lengthened (positively chirped) to about 200 ps, everything else being identical, the diameter at 100 m was more than three times larger than the initial diameter. Similar observation was obtained recently by Béjot et al. (2007) when they propagated a 32 TW, 26 J, 1053 nm, 570 fs laser pulse vertically into the

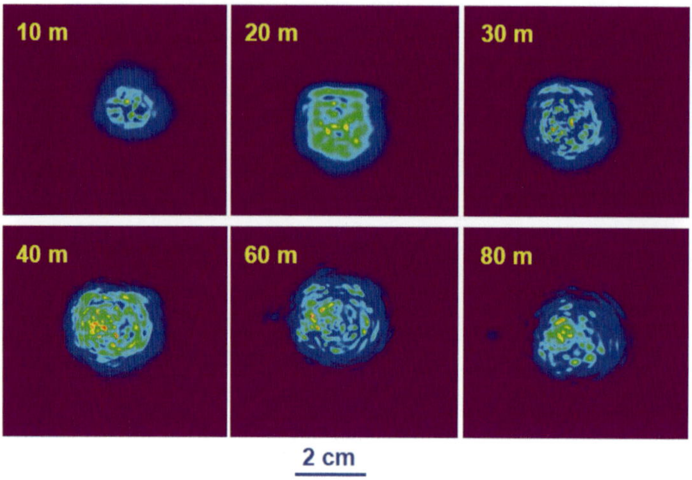

Fig. 4.3 The beam pattern at the fundamental wavelength of the laser pulse at different distances from the last mirror. The pictures are taken by the CCD camera with a band pass filter of 800 ± 20 nm. The reservoir is essentially the whole (weak) background pattern while the hot spots are evident. Note that the pattern does not diverge too much

atmosphere. The reason for the quasi-maintenance of the beam diameter is due to slice-by-slice self-focusing. Each self-focusing of a "warm" zone means that the energy of the pulse is "sucked" toward the self-focal region and then it is released back to the background reservoir after the self-focus. Successive slices would repeat the same process. When there are many self-foci (multiple filamentation), each of the self-foci contributes to "sucking" energy from the background reservoir toward their respective focal zone and releasing it back to the reservoir. So long as the power in a "warm" zone is above the critical power for self-focusing, this "sucking" process would take place and slow down the linear diffraction of the pulse. We call this pulse a "self-guided hot light pulse" (Liu et al., 2004). This is perhaps another indication that the pulse could propagate very far in air and still might give rise to high intensities in the pulse. However, the high clamped intensity in air, 5×10^{13}W/cm^2, is still yet to be proven directly in a long distance (more than a few hundred meters) field experiment. It is a challenge to experimentalists since the beginning of extensive research on filamentation in air a decade ago (see below for more discussion).

4.3 Interference and Competition of Multiple Filaments

As mentioned above, the hot/warm zones are not independent from one another because they all try to feed energy from the whole pulse's back ground reservoir into their own self-foci. This constitutes a competition for energy (Mlejnek et al., 1999; Hosseini et al., 2003, 2004; Luo et al., 2005). The underlying physics of multiple filament competition is essentially field redistribution inside the pulse during propagation. It consists of two inter-related scenarios. One scenario is linear field interference inside the pulse during propagation and self-deformation through the optical medium. The other is nonlinear field redistribution due to nonlinear propagation effects.

In the first scenario, consider first a "warm" zone as being a single pulse. When a slice of the "warm" zone self-focuses toward a high intensity spot, tunnel ionization of air molecules occurs and the intensity is clamped. With further propagation, the slice will diverge, becoming a conical wave which interferes with the background field (a quasi-plane wave) giving rise to concentric rings around the self-focus (Aközbek et al., 2002b; Chin et al., 2002). When two adjacent "warm" zones self-focus into two nearby self-foci, the two sets of rings (or rather, the two conical waves and the background quasi-plane wave) will interfere giving rise to a star-like pattern (Chin et al., 2002). This is shown in Fig. 2.19 in which two filaments/hot spots (one strong and one weak) interfere forming such a star-like structure. Calculations based upon the above idea of the interference of conical waves and the back ground quasi-plane wave indeed reproduce the observations (Chin et al., 2002). When more than two nearby filaments interfere, the resultant field would give rise to more complicated structures with more new "warm" zones which would undergo self-focusing again during further propagation. New "children" filaments are thus formed at new positions both along the propagation axis and on the cross sectional surface (Hosseini et al., 2004; Luo et al., 2005).

However, if the "warm" zones or filaments are far apart, interference would be too weak to form new and sufficiently "warm" zones for self-focusing. This would constitute the second scenario in which the initial "warm" zones would each go their own way as if they were independent. During the propagation, the nonlinear self-focusing effect would help each of the "warm" zones to "pull" the field toward its own self-focus as if each filament "sucks" energy from the background reservoir. The consequence of this competition for energy from the same background reservoir would be such that the filaments do not have enough energy to develop fully into self-steepening or maturity.

Such competition for energy would also take place in the first scenario with filaments adjacent to one another but in a more constructive way because the filaments are close to one another. Apart from creating "children" filaments, the central bunch of filaments would collectively "suck" energy from the background reservoir toward them as if they were one single filament. The consequence of this latter case is that all these filaments become mature almost at the same time over a short distance of propagation. In air, in such a short distance, the nitrogen fluorescence signal is very strong. After this short distance, the children filaments take over but will not be as strong as before.

These two scenarios were observed in our recent experiments (Hosseini et al., 2004; Luo et al., 2005). We measured the nitrogen back scattered fluorescence from long filaments in air using a LIDAR (laser radar) technique. We found that with a beam diameter of about 25 mm ($1/e^2$ of fluence) over which the multiple "warm" zones were sufficiently far apart, the back scattered fluorescence from the generated

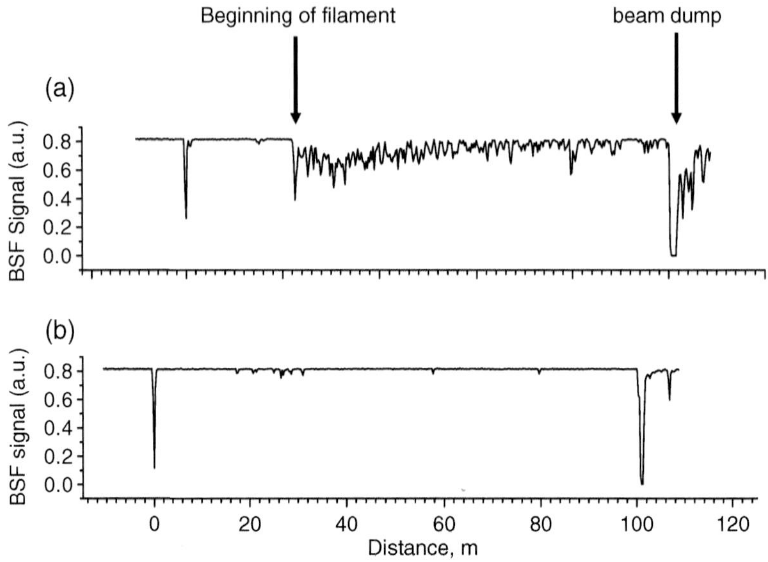

Fig. 4.4 Example of the pulse to pulse difference in the back scattered fluorescence (BSF) in air as a function of distance. The fluorescence of nitrogen was detected using a photomultiplier and a band pass filter (UG11). Both (**a**) and (**b**) are under identical laser conditions: 800 nm, 46 mJ, 42 fs

multiple filaments had a huge fluctuation. For the same input laser peak power of about 1 Terawatt (TW, or 10^{12} W), the fluorescence signal sometimes came from the full propagation length of 100 m (Fig. 4.4a), but sometimes, there was very little after about 20 m of filamentation and sometimes, nothing at all (Fig. 4.4b). However, if the beam diameter is made three times smaller while keeping the distribution of "warm" zones roughly the same; i.e., forcing the generated hot spots to be closed to one another, we could detect back-scattered fluorescence from all laser shots; in the first 10 m, the fluorescence intensity was more than 100 times stronger than the case of larger diameter (25 mm) beam (Luo et al., 2005). However, the fluorescence signal from beyond the first 10 m range becomes very weak even using a photomultiplier to detect the signal. Changing the divergence and the chirp of the pulse would not move this strong filament zone significantly. This is because the diameter of each of the hot zones in the small diameter pulse becomes so small that this hot zone will self-focus at a short distance (Eq. 2.1). Calculations using the same formulation as discussed in the case of a single filament but with an input beam pattern consisting of two or more "warm" zones confirmed these observations (Hosseini et al., 2004; Kosareva et al., 2006).

Other groups have calculated what they called either "optical turbulence" effect (Mlejnek et al., 1999) or "filament fusion and breakup" (Tzortzakis et al., 2001; Bergé et al., 2004a, b) in air through the solution of nonlinear equations. Their 3D numerical results were very similar; i.e., there were many high intensity narrow cylindrical zones (multiple filaments) randomly distributed along the propagation direction. Mlejnek et al. (1999) interpreted this phenomenon as an optical turbulence effect because they started with a random perturbation across the laser beam pattern and obtained randomly changing filamentation. Tzortzakis et al. (2001) and Bergé et al. (2004a, b) interpreted this as obvious filament fusion and breakup. These interpretations seemed to be based upon the natural "look" of those filament zones that meet or separate randomly. We believe the physics behind all these is the same as what we analyzed above; i.e., filament competition through either field interference effect or energy competition or both.

Another important yet often neglected origin of multiple filamentation in a real experiment is the non-cylindrical symmetry of the beam cross section; i.e., a non-circular symmetry of the beam pattern also gives rise to multiple filamentation. A high power Ti-sapphire femtosecond laser pulse generated by a non-cylindrically symmetrical technique using cross beam pumping of the amplifier (a popular method) will certainly give rise to such an effect if spatial filtering is not employed. However, spatial filtering would seldom work at the TW level because the pin-hole that does the filtering would often be damaged even in vacuum. Under such a situation, the beam pattern is elliptical in shape. The ellipticity depends on the detailed technique of beam alignment. When such a beam pattern propagates and self-focuses in air, a discontinuous ring that is elliptical in shape would be generated. Several hot zones occur around the elliptical ring. Multiple filamentation could occur from these hot zones if the energy content of the pulse is high enough. The physical reason for creating the hot zones around the elliptical ring is the following.

A slice of such a pulse with elliptical cross section would self-focus toward a line focus where ionization and intensity clamping takes place. This would create first of all a line of self-foci (filaments) in practice because the intensity along the line is not uniform. Subsequent divergence of this line source gives rise to an oval-shaped conical wave. Interference with the quasi-plane wave of the background field would result in discontinuous elliptical rings. The discontinuity around the elliptical ring is due to the fact that the phase difference between the conical field and the quasi-plane wave field changes around the ellipse. Hence, there will be constructive and destructive interferences. The pattern of hot zones around the ellipse would be symmetrical though. Such a phenomenon was recently predicted numerically (Dubietis et al., 2004a). Such kind of hot zones along a line perpendicular to the propagation axis can also be generated if we use a focusing lens and induce astigmatism. At the focal zone, line focusing due to the astigmatism takes place (Kamali et al., 2009a) and the same type of discontinuous elliptical pattern would occur which would degenerate into multiple filaments (Carrasco et al., 2003).

4.4 Theory of Multiple Filamentation

The theory of multiple filamentation is much more complicated because the initial intensity distribution across the wave front of a real experimental pulse with power much higher than the critical power cannot be represented by any ideal mathematical function such as the usual one with cylindrical symmetry. In fact, the transverse intensity distribution of a pulse coming out from a normal laser system that would give rise to multiple filamentation is random not only from system to system but also from shot to shot. That is to say, the theory has to be calculated on a case-by-case basis. There have been attempts to regularize the transverse distribution using regular structures to force a certain wave front distribution before propagation (Kandidov et al., 2005; Schroeder and Chin, 2004). Readers interested in the current theories of multiple filamentation are referred to the following papers: Kosareva et al. (2006), Kandidov et al. (2005), Hosseini et al. (2004), Mlejnek et al. (1999), Tzortzakis et al. (2001), Bergé et al. (2004a, b).

4.5 The Challenge of Long Distance Filamentation

One of the most attractive features for application of femtosecond Ti-sapphire laser pulses is the possibility in principle of filamentation at long distance in air. The feasibility of remote sensing and control using practically one laser only would have an enormous impact. However, current laser technology is still limiting us to create filaments in air at relatively short distances, up to the order of a hundred meters or so. The limiting factor is related to the beam quality, not the peak power. Almost all currently available multi-TW level femtosecond Ti-sapphire laser systems, both commercial and home-made, give out hat-top pulses; i.e., pulses whose intensity distribution across the cross section of the beam is almost flat with a lot of fluctuation around the rim of the hat-top. Such intensity fluctuation will give rise to multiple filaments many of which are around the perimeter of the cross sectional surface (Méchain et al., 2004). Moreover, the beam diameter is large, of the order of many cm or more. Such quality would give rise to filament competition for energy inside the reservoir. The negative consequence is evident, as we have discussed above. That is to say, it is very difficult to create high intensity filaments at long distances. So far, there is no direct measurement showing that the high clamped intensity in air (5×10^{13} W/cm^2) could be achieved at distances beyond a few hundred meters. Yet, this is the desirable intensity for many applications at remote position such as pollutant measurement, target identification, lightning control, etc. One near exception is the work by Béjot et al. (2007) in which they claimed to have created intense filament vertically in air that ends at about 350 m from the laser source that sends out pulses at 32 TW, 26 J/1053 nm/570 fs. The intensity was slightly higher than 5×10^{13} W/cm^2 because they used a longer wavelength laser rather than the popular 800 nm pulses.

4.6 Long Distance Multiple Filamentation Control

We have discussed how a small diameter beam would lead to more constructive interference among the various filaments close to one another, resulting in a strong filament bunch, but only through a short distance. To overcome this difficulty, we designed a focusing telescopic technique. The beam diameter is first enlarged by a concave lens (convex mirror) so that the diameter of each of the hot zones inside the pulse is enlarged. The principle is that such enlarged hot zones would have their self-focusing distance z_f (cf. Eq. 2.1) lengthened. A large diameter focusing lens (concave mirror) then focuses the pulse at a distance that depends on the distance between the diverging and focusing lenses (mirrors). The idea is to control the distance between the two lenses (mirrors) such that the self-focusing distance is much longer than the external focusing length. Under this condition, filaments will mostly occur before the geometrical focus (see Section 2.3.4). Since the beam diameter is now relatively smaller in the focal region, the hot zones are also forced to be closer to one another. This would enhance constructive interference among the filaments and hence induce a large number of mature filaments to occur. This technique was put to experimental test successfully by Liu et al. (2006b) and was applied to enhance significantly the detection efficiency of water aerosols containing metallic oxides (Daigle et al., 2007a, b).

Ideally speaking, we need to have a smooth wave front with an intensity distribution such as a Gaussian distribution. But in practice, the beam is far from being Gaussian. Even if one starts with a Gaussian beam, atmospheric conditions such as turbulence, temperature gradient, etc. would deteriorate the beam quality in such a way that there are many hot/warm zones across the beam cross section. Each of these zones would self-focus into a filament but they all compete for the energy inside the reservoir. The result is that it is almost impossible to generate strong filaments even at distances of less than 100 m.

The author has experienced this "negative" effect in two major laboratories with multi-TW capabilities although no quantitative measurement was done yet. One is with the transportable 5 TW laser system inside a container at the DRDC-Valcartier laboratory in Valcartier, Quebec, Canada. Negatively chirped pulses at a few ps at more than 200 mJ per pulse could not create intense filament even at 60 m away in the field in the cold winter (–10 to –20°C). Once the heating around the container is turned on, many more randomly distributed warm zones occur in the large elliptical beam pattern (22 × 28 mm at $1/e^2$ of the fluence distribution) than when the heating is off. Using a focusing telescopic system as described above for the control of multiple filaments, the beam diameter at the focal zone at a few tens of meters away in the cold open field is roughly twice as large as compared to the case when the heating system is turned off. Some of the remote sensing results under such severe conditions are described in Kamali et al, 2009b.

The other experience is at the Shanghai Institute of Optics and Fine Mechanics of the Chinese Academy of Sciences, Shanghai, China. The hat-top 10 cm-diameter pulse at 100 TW but with a lot of hot zones around the perimeter of the beam cross section deteriorates quickly at a distance of less than 30 m. No systematic long distance experiment could be carried out yet with this system. (Even so, an important experiment was performed. Short distance propagation by focusing the beam to about 10 m or so in air shows that the peak intensity inside the multiple filaments can never increase significantly due to intensity clamping (Kosareva et al, 2009). In chapter 8, a little more detail is given.)

To overcome this inherent difficulty arising from the laser design, it is necessary to improve either the beam quality using adaptive optics or through better laser design. Adaptive optics cannot improve the beam quality of a hat-top pulse easily. However, some advancement using a deformable mirror has been realized in the author's laboratory recently. The back scattered fluorescence signal from air was improved significantly (Daigle et al., 2008).

Chapter 5
Filamentation Nonlinear Optics: General

5.1 Self-Actions

Self-focusing and filamentation of a powerful femtosecond laser pulse in an optical medium result in the following self-actions. These self-actions were almost totally unexpected based upon the normal wisdom of the propagation of a long nanosecond laser pulse. These self-actions include

1. Self-avoidance of breakdown in air;
2. Self-stabilized high intensity in the filament core;
3. Self-transformation into a white light chirped laser pulse;
4. Self-spatial filtering;
5. Self-remote projection in air;
6. Self-pulse compression;
7. Self-group-phase locking of other pulses inside the filament.

The first self-action on the avoidance of breakdown in air is discussed in Chapter 1. The second self-action is essentially intensity clamping discussed in Chapter 2. The third self-action, namely self-transformation into a white light laser is also discussed in Chapter 2. So is self-spatial filtering, the fourth self-action. The last self-action, namely self-group-phase locking, will be discussed separately in Chapter 6. The next two sections give a description of self-actions 5 and 6.

All these self-actions are happening inside the filament core due to the filamentation process and all the physical processes governing these actions are nonlinear. In particular, because of the high intensity inside the filament core, many efficient nonlinear interactions such as third harmonic generation, four-wave-mixing, waveguide writing in glasses, etc. could be induced inside the core. We call these self-actions and the many interactions, known and to be discovered, inside the filament core or due to filamentation "filamentation nonlinear optics" (Chin et al., 2007).

S.L. Chin, *Femtosecond Laser Filamentation*, Springer Series on Atomic, Optical, and Plasma Physics 55, DOI 10.1007/978-1-4419-0688-5_5,

5.2 Self-Remote Projection in Air

We assume a pulse whose intensity distribution is Gaussian in space (i.e., across the beam profile) and in time. The pulse duration is of the order of a few tens of fs to a few hundred fs. Because of the slice-by-slice self-focusing process, a filament is produced. The length of the filament during the free propagation of such an intense femtosecond laser pulse follows the evolution of the filament as described in Chapter 2. The beginning of the filament is, in principle, governed by the following equation which is a repetition of Eq. (2.1) for the sake of convenience to the readers.

$$z_f = \frac{0.367ka_0^2}{\left\{\left[\left({}^{P}\!/\!{}_{P_c}\right)^{1/2} - 0.852\right]^2 - 0.0219\right\}^{1/2}} \tag{5.1}$$

a_0 is the radius at $1/e$ of the spatial Gaussian intensity distribution. From this Eq. (5.1), we can see the following consequences. The self-focusing distance, z_f, varies as a function of the beam radius, a_0, and the ratio of the peak power over the critical power for self-focusing, P/P_c. The filament ends when P/P_c is unity; i.e., the peak power equals the critical power when the self-focusing position is at infinity by definition of the critical power (see Chapter 2). Thus, for a given peak power, the position of the beginning of the filament is the self-focusing distance of the most powerful slice of the pulse (peak power position). The higher the peak power is, the shorter is the self-focusing distance and the longer is the filament.

To move the filament to a longer distance, it is sufficient to know how to move the beginning of the filament. In the above case, one can increase the radius of the beam using a telescope while keeping the peak power (or the energy and the duration) of the pulse constant. (One might say that in so doing, the intensity distribution across the beam is changed. This would decrease the Kerr self-focusing effect of the wave front which in turn would influence z_f. However, this consideration is not correct because Eq. (5.1) has already taken this into account during its derivation with a_0 being the radius of the spatial Gaussian intensity distribution.)

One can further increase z_f by using chirped pulses (both negatively and positively). Chirping the pulses would have an effect of reducing the peak power. Hence, for a constant beam radius, P/P_c becomes smaller; i.e., z_f is longer. Figure 5.1 shows an experimental result when the pulse is chirped either negatively or positively by changing the distance between the grating pair.

Using an optical element (lens, mirror, etc.) to create a diverging wave front could also increase the self-focusing length (see Section 2.3.4) through the following equation

$$1/z_f + 1/f = 1/z_f' \tag{5.2}$$

where f is the focal length of the optical element which could be positive or negative and z_f' is the effective self-focusing length.

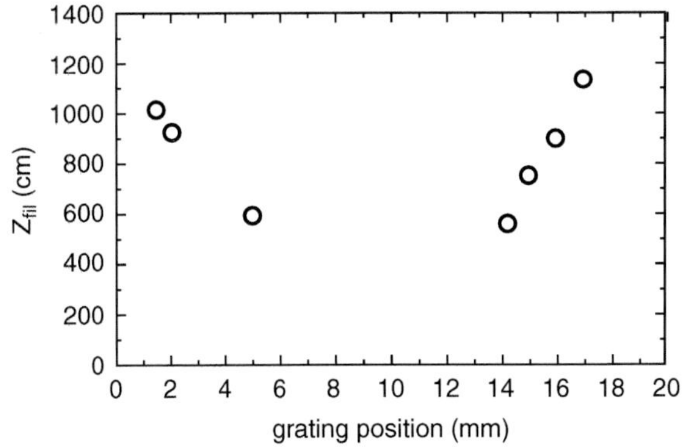

Fig. 5.1 The beginning of filamentation in air (i.e., self-focusing distance of the peak of the pulse) as a function of positive and negative chirps of the pulse; i.e., as a function of the spacing between the parallel grating pair in the compressor. z_{fil} is the distance from the output of the compressor to the beginning of the filament measured by burn paper. Input beam radius at $1/e$, intensity level $a_0 = 2.1$ mm, pulse energy: 6.2 mJ; optimum pulse duration: 250 fs. Positive chirp is at larger grating positions (towards the right hand side of the minimum); negative chirp towards opposite positions

In the case of free propagation of the laser pulse with multiple filamentation, we could assume that each hot zone represents a Gaussian beam which will self-focus by itself. Its self-focusing distance is effectively the z_f given by Eq. (5.1). We again call this effective self-focusing length z_f'. At this distance, there is at least a high intensity spot with plasma generation inside the pulse. Since the behavior of each filament is, in principle, similar, we would expect that the effective self-focusing length z_f' in the case of multiple filamentation is similar to that in single filamentation. However, because of interference of the fields and the non-uniformity of intensity distribution and curvature of the wave front, different filaments are born at different positions of propagation. Filament competition is a negative effect that will limit severely the formation of strong ionizing filaments if the beam cross section is large (see Chapter 4). Constructive interference of the fields from filaments closed to one another in a small diameter beam (Chapter 4) would quickly degenerate into many small scale hot spots whose individual z_f is short because of the small radius of the hot spots. This would reduce and hence limit the effective distance z_f'.

From the above discussion and at the current level of understanding, the best way to increase the self-focusing length is to operate in a single filament regime using a Gaussian pulse with a large beam radius. In practice, this is very difficult to achieve because even if one could obtain a very good Gaussian pulse, propagation will deteriorate the beam quality so that it will degenerate into multiple filamentation sooner or later.

An alternative idea is to use a deformable mirror to correct the wave front. This is a current hot subject of research being carried out in many laboratories including that of the author where partial success has been achieved recently (Daigle et al., 2008).

5.3 Self-Pulse Compression

As shown in the series of figures in Chapter 3, during the filamentation of a femtosecond laser pulse, at first the back part of the pulse is distorted and decreased in intensity due to the interaction with the plasma left behind by the self-focusing slices at the front part of the pulse. This results in the front part of the pulse that keeps on self-focusing; i.e., the pulse is effectively shortened. Meanwhile, the distorted energy at the back part of the pulse would self-focus again after a certain propagation distance leading to the formation of two peaks; i.e., pulse splitting (Fig. 2.24). With further propagation of the pulse, the front part of the pulse decreases while the back part still persists (Fig. 2.24) till the end of the filamentation.

The first detailed experimental observation of self-pulse compression of a Ti-sapphire laser pulse in atomic, molecular gases as well as in solids was done in Riken in Japan (Koprinkov et al., 2000). Up to 10 times compression down to 30 fs was obtained. Later, starting from 43 fs, Hauri et al (2004) was able to obtain 5.7 fs pulses after self-compression. This opened up a new door for the compression of the pulse down to single cycle level. Many laboratories jumped into the field immediately after, since this is a very simple technique that requires almost no effort to do the compression. One of the major reasons that these laboratories are interested in generating an intense single cycle pulse at the Ti-sapphire laser wavelength (800 nm) is to generate high order harmonics in gases so as to generate, in turn, a single attosecond (10^{-18} s) pulse.

However, there are inherent difficulties. The pulse is never "clean" in principle. One cannot extract the pulse at the end of the filament because there will be almost no energy in the filament core (see Fig. 2.24) and the temporal form of the pulse is not good at all (multiple peaks). The background reservoir is a major problem to overcome because it would constitute a background noise both spatially and temporally. Since the background reservoir contains a lot of the total pulse energy, even if one can use a pinhole to block the reservoir spatially at a certain propagating position, the central core still contains a background with a long duration in the temporal dimension. The temporal distribution of the reservoir is perhaps most difficult to filter out.

A compromise to obtain a "clean" pulse is to extract the pulse near the end of the efficiently ionizing filament zone before pulse splitting (Kosareva et al., 2007, 2008). This would give an optimized self-compressed pulse which contains one rather clean peak with high enough energy but not the shortest pulse.

5.4 Exploitations of the Self-Actions

One of the naturally attractive self-actions is the self projection of the intense filament (core) to a long distance in air. No other long pulse laser could do so. One could thus do a lot of things by projecting the filament onto a distant target. For example, one can remotely detect this target. If it is a solid target, a "clean" but intense plasma is generated on the surface and its distinct fluorescence can be measured by a LIDAR (see Chapter 7). Similarly, a gas or aerosol at a distance could be fragmented by the high intensity in the filament core and emit "clean" fluorescence without breakdown (self-avoidance of breakdown) that could be detected by a LIDAR (Chapter 7). As demonstrated in Chapter 7, one could, in principle, detect all targets at a distance using only one laser.

The pulse, through self-compression, self-phase-modulation and self-steepening, self-transforms into a white light laser pulse. This white light source in the sky could be used to detect various gaseous targets through absorption of the back-scattered white light from cloud or air (Chapter 7). From the defense or safety point of view, the white light could also be used to jam/saturate/even damage a light detector (including the eyes of humans) at a distance. The high intensity core, even after filamentation with a lower intensity in air (Chapter 2), could generate white light inside a distant target's window because the critical power for self-focusing in condensed media is only a few megawatts. Let's take the experimental example in Section 2.6. The intensity after filamentation is estimated to be 10^{11} W/cm^2. Assuming that the diameter is 1 mm, the power is of the order of 10^8–10^9 W. This is much higher than the critical power in condensed media. This secondary but intense white light thus generated inside the window would also jam/saturate/even damage the detectors behind the window.

Intensity clamping is a profound phenomenon as mentioned in Section 2.3.2. Moreover, the high intensity filament core has an excellent spatial quality because of self-spatial filtering. Thus, one can make use of this quality to write waveguides in glass, or to induce any reaction such as chemical reaction whose rate would be rather constant.

The ionizing filament can also act as a conductor to guide electrical discharge (including lightning control). If one applies a high voltage between the extremities of this column, a guided discharge is possible. The physics, though, is not what we normally would expect; i.e., it is apparently not the charged column that conducts the electricity. This is probably because the electron density is low and is not uniform along the filament. According to Tzortzakis et al. (2001), it is due to electron-ion recombination inside the weak plasma column which heats up the central part. The thermal expansion of the heated column leads to a reduction of pressure. The high voltage at the extremities of the plasma column would then discharge through the low pressure channel. Such discharge mechanisms have been first investigated by La Fontaine et al (1999) and Vidal et al. (2000) where the ideas of density lowering and non-uniformity are clearly expressed. Recent attempts to control lightning in laboratory tests were positive and the future success of this idea seems very promising (Rodriguez et al., 2002; Pépin et al., 2001; Comtois et al., 2003).

An experiment by the Teramobile group (Kasparian et al., 2003) shows that filamentation inside a cloud chamber can induce nucleation of the supersaturated water vapor. The physics is similar to the standard nuclear physics experiment in which ion tracks are left behind by gamma rays, for example, and water molecules would nucleate around such ions resulting in water droplets formation. A simple reasoning is that a water molecule has a strong permanent dipole moment; hence any charge particle will attract many water dipoles around it forming a droplet eventually. Ionization inside the filaments would have similar consequence. We could even stretch our imagination and ask if this observation could be applied to artificial rain making by generating filaments inside a rain cloud containing supersaturated water vapor.

The ionizing filament, being a plasma column, could act as an antenna to guide radio waves (Dormidonov et al., 2007). This is recently proven to be feasible (Châteauneuf et al., 2008).

Many more applications remain to be discovered when more scientists go into this new field.

Chapter 6
Filamentation Nonlinear Optics: Third Harmonic Generation and Four-Wave-Mixing Inside a Filament

6.1 Introduction

Aközbek et al. (2002a) discovered a new phenomenon called self-phase locking both experimentally and theoretically during third harmonic generation inside a filament in air. During the filamentation of an ultrashort and intense laser pulse in air, the generated third harmonic pulse inside the filament propagates with the fundamental at the same group velocity due to nonlinear intensity dependent cross interaction between the two pulses. The cross interaction through the Kerr effect generates what the authors (Aközbek et al., 2002a) called two-color filament at which the relative phase between the fundamental and the third harmonic pulses is self-locked together despite the group velocity dispersion between the fundamental and the third harmonic. We call this self-group-phase locking. This nonlinear phase locking phenomenon could be extended to other nonlinear frequency generation processes such as four-wave-mixing inside the filament resulting in phase matching of all the pulses of different frequencies over long filamentation distances. This chapter discusses the generation of third harmonic in air as well as four-wave-mixing (4WM) inside the filament in air by mixing a tunable infrared (IR) pulse with the filament of an 800 nm pump pulse. Tunable few cycle visible pulses of almost the highest beam quality were observed. Projection to tera-hertz generation inside the filament as well as the generation of other frequencies through 4WM will also be briefly discussed.

6.2 Third Harmonic Generation Inside a Filament in Air (Theoretical Analysis)

If we assume that a laser beam propagates and undergoes filamentation in air, it would naturally generate third harmonic (TH) (Boyd, 2003). Inside the filament, because of the high intensity, it is expected that the efficiency of TH generation (THG) will be high. However, in normal long pulse nonlinear optics, if one uses a lens to focus the laser pulse in air, the third harmonic generated before the focus of a Gaussian beam will interfere destructively with that generated after the focus

S.L. Chin, *Femtosecond Laser Filamentation*, Springer Series on Atomic, Optical, and Plasma Physics 55, DOI 10.1007/978-1-4419-0688-5_6,

because of the Gouy phase shift[1] of π of the pump pulse across the focal plane of a Gaussian beam (Boyd, 2003, pp. 93–97). Thus the efficiency should be very weak. However, during filamentation, self-focusing starts before the geometrical focusing and ends practically at the geometrical focus. There is no Gouy phase shift. Thus, the efficiency of the TH generation is expected to be high. What is not expected is that the TH pulse in the filament propagates at practically the same speed as that of the pump pulse without any dispersion. (Note that in linear optics, the TH will propagate slower than the pump in air because the linear index of refraction of the TH is higher than that of the pump at 800 nm.) This is because the nonlinear polarization that governs the speed of propagation/index of refraction of the pulses contains terms involving the crossing of the fundamental field and the TH field. The index of refraction of the pump pulse is practically immune from the influence of the TH field because the latter is much weaker than the former. In fact, the index of refraction of the fundamental pulse is the normal $n = n_0 + n_2 I_\omega$, where n, n_0, $n_2 I_\omega$ are the indices of refraction, total, linear and nonlinear; I_ω being the intensity at the pump inside the filament core which is very high and practically constant because of intensity clamping. The index of refraction of the TH pulse contains a non-negligible cross term that depends on the intensity of the pump as well as a term depending purely on the intensity of the pump. Since the pump intensity is high, the index of refraction of the TH is practically controlled by the pump. Consequently, the TH pulse is dragged along by the pump pulse.

Further analysis (see below) by the author based upon the phase matching condition shows that the TH pulse, though propagating at the same speed as the pump, stays faithfully behind the pump pulse by a distance of half the wavelength of the TH or 1/6th of the fundamental wavelength. In the temporal scale, the TH pulse stays at about $\lambda/(6c) \cong 0.44$ fs behind the pump pulse. Literally, it is as if the pump pulse at the front would prepare the path for the TH pulse to follow. Physically, this seems to mean that the electronic nonlinear response of air peaks at 0.44 fs.

The theory of third harmonic generation is to set up the wave equations corresponding to the simultaneous propagation of the fundamental and TH waves. This is done by solving the Maxwell's equations. A detailed derivation and analysis of the theory can be found in Aközbek et al. (2002a, 2003) and Théberge (2007). In this chapter, the physical idea together with some essential equations of the theory will be discussed. The general wave equation in the time domain under slowly varying envelop approximation is given by (for detail, see Aközbek et al., 2002a, 2003; Théberge, 2007. A brief explanation of the derivation is given at the end of this chapter)

---

[1] During harmonic generation, for example the third harmonic, using a Gaussian beam, the conversion is identically zero when the beam is tightly focused and when the phase matching condition is zero or negative. This phenomenon is essentially due to the phase change of π when crossing the focal plane of a Gaussian beam and is first investigated in detail by Gouy (Boyd, 2003, pp. 93–97).

$$\Delta_{\perp}A + 2ik_0\left[\frac{\partial A}{\partial z} + \frac{1}{v_g}\frac{\partial A}{\partial t}\right] - k_0 k\frac{\partial^2 A}{\partial t^2} - \frac{\omega_p^2}{c^2}A + i\alpha^2 A = -\frac{k_0^2}{\varepsilon_0}P^{NL}e^{-ik_0 z} \quad (6.1)$$

where $A(r,z,t)$ is the amplitude function of the electric field $E(r,z,t) = A(r,z,t)\,e^{-i(\omega_0 t - k_0 z)} + c.c.$ $A(r,z,t)$ could be the fundamental field $A_\omega(r,z,t)$ or the TH field $A_{3\omega}(r,z,t)$ and v_g, the corresponding group velocity. The plasma angular frequency is given by $\omega_p = \sqrt{N_e e^2/\varepsilon_0 m_e}$. Multiphoton/tunnel ionization of air molecules gives rise to this plasma whose electron density is N_e. The relation between k_0, the linear wave vector and k, the wave vector is given by the following equation:

$$k = k_0 + (\omega - \omega_0)\,k' + \frac{1}{2}(\omega - \omega_0)^2\,k'' + \cdots \quad (6.2)$$

k_0 is given by $k_0 = \frac{\omega}{c}\sqrt{1+\chi^{(1)'}}$ where $\chi(1)'$ is the real part of the linear susceptibility of air $\chi^{(1)} = \chi^{(1)'} - \left(\omega_p^2/\omega^2\right) + i\chi^{(1)''}$ and $\alpha = (\omega/c)\sqrt{\chi^{(1)''}}$ corresponds to the absorption coefficient. The nonlinear polarization induced in the material (air), P^{NL}, in the case of THG is given by (neglecting higher order terms) (Théberge, 2007):

$$\begin{aligned} P^{(3)} &= \varepsilon_0\chi^{(3)}EEE \\ E &\equiv E(r,z,t);\; P \equiv P(r,z,t) \end{aligned} \quad (6.3)$$

where

$$E(r,z,t) = \frac{1}{2}A_\omega(r,z,t)\,e^{i(k_\omega z - \omega_0 t)} + \frac{1}{2}A_{3\omega}(r,z,t)\,e^{i(k_{3\omega}z - 3\omega_0 t)} + c.c. \quad (6.4)$$

Here, we have assumed that the linear polarizations of the pump and the TH are parallel; hence, scalar terms are used. Equation (6.3), in view of Eq. (6.4), contains many cross terms. Apart from the nonlinear material polarizations $P_\omega^{(3)}$ and $P_{3\omega}^{(3)}$ oscillating at the frequencies ω and 3ω, other nonlinear material polarizations oscillating at higher frequencies also exist, e.g., 5ω through four-wave-mixing (4WM) of two TH photons and one fundamental photon or THG of the TH photons. We shall neglect these higher order (hence, less probable) material polarization terms and consider only those nonlinear material polarization terms pertaining to interactions giving rise to the TH and the fundamental frequencies (Théberge, 2007).

$$P_{3\omega}^{(3)} = \frac{1}{8}\varepsilon_0\chi^{(3)}A_\omega^3 e^{i(3k_w z - 3\omega_0 t)} \text{ (THG of the fundamental } A_\omega) \quad (6.5)$$

$$P_\omega^{(3)} = \frac{3}{8}\varepsilon_0\chi^{(3)}\,|A_\omega|^2 A_\omega e^{i(k_\omega z - \omega_0 t)} \text{ (SPM of the fundamental } A_\omega) \quad (6.6)$$

$$P_\omega^{(3)} = \frac{3}{4}\varepsilon_0\chi^{(3)}\,|A_{3\omega}|^2 A_\omega e^{i(k_\omega z - \omega_0 t)} \text{ (Cross phase modulation (XPM), influence of TH on } A_\omega) \quad (6.7)$$

$$P_{3\omega}^{(3)} = \frac{3}{8}\varepsilon_0 \chi^{(3)} |A_{3\omega}|^2 A_{3\omega} e^{i(k_{3\omega}z - 3\omega_0 t)} \text{ (SPM of the TH } A_{3\omega}) \tag{6.8}$$

$$P_{3\omega}^{(3)} = \frac{3}{4}\varepsilon_0 \chi^{(3)} |A_{\omega}|^2 A_{3\omega} e^{i(k_{3\omega}z - 3\omega_0 t)} \text{ (XPM, influence of the fundamental on the } A_{3\omega}) \tag{6.9}$$

$$P_{\omega}^{(3)} = \frac{3}{8}\varepsilon_0 \chi^{(3)} A_{\omega}^{*2} A_{3\omega} e^{i((k_{3\omega} - 2k_w)z - \omega_0 t)} \text{ (4WM: } \omega_{4\mathrm{WM}} = \omega_{\mathrm{TH}} - 2\omega_{\mathrm{fundamental}} = \omega) \tag{6.10}$$

We recall that the index of refraction is related to the polarization of the medium by the relation (using scalar field)

$$n = \sqrt{\varepsilon/\varepsilon_0} = \sqrt{1+\chi} = \sqrt{1 + P/E} \tag{6.11}$$

where n, ϵ and χ are the index of refraction, the dielectric constant and the total susceptibility of the medium, respectively. P and E are the total scalar material polarization and the electric field at the appropriate frequency, respectively. Thus, the larger the material polarization is, the larger the index of refraction. Imagine a pulse with a Gaussian distribution of intensity across the wave front. Equations (6.6) and (6.8) give rise to self-focusing. After substituting Eqs. (6.4), (6.5), (6.6), (6.7), (6.8), (6.9), and (6.10) into (6.1), and taking into account the plasma generation process due to multiphoton/tunnel ionization, one obtains (for detail, see Aközbek et al., 2002a, 2003; Théberge, 2007) the following coupled equations for the fundamental (Eq. 6.12) and the TH (Eq. 6.13):

$$\begin{aligned} &\left[\Delta_\perp + 2ik_{0,\omega}\frac{\partial}{\partial z} - k_{0,\omega}\, k''\big|_{\omega}\frac{\partial^2}{\partial t^2} - \frac{\omega_p^2}{c^2} + i\alpha_\omega^2\right]A_\omega \\ &= -k_{0,\omega}^2 \frac{3}{4}\chi^{(3)}\left[|A_\omega|^2 A_\omega + 2|A_{3\omega}|^2 A_\omega + A_\omega^{*2} A_{3\omega} e^{-i(\Delta k)z}\right] \end{aligned} \tag{6.12}$$

$$\begin{aligned} &\left[\Delta_\perp + 2ik_{0,3\omega}\frac{\partial}{\partial z} - 2k_{0,3\omega}\Delta k + 2ik_{0,3\omega}\frac{1}{\Delta v_g}\frac{\partial}{\partial t} - k_{0,3\omega}\, k''\big|_{3\omega}\frac{\partial^2}{\partial t^2} - \frac{\omega_p^2}{c^2} + i\alpha_{3\omega}^2\right]A_{3\omega} \\ &= -k_{0,3\omega}^2 \frac{3}{4}\chi^{(3)}\left[|A_{3\omega}|^2 A_{3\omega} + 2|A_\omega|^2 A_{3\omega} + \frac{A_\omega^3}{3}\right] \end{aligned} \tag{6.13}$$

where the linear wave vectors are $k_{0,\omega} = n_{0,\omega}k_0$ and $k_{0,3\omega} = n_{0,3\omega}3k_0$ and the wave vector mismatch is given by $\Delta k = 3k_0\left(n_{0,\omega} - n_{0,3\omega}\right)$. $\left(1/\Delta v_g\right) = \left(v_g(3\omega)\right)^{-1} - \left(v_g(\omega)\right)^{-1}$ represents the group velocity mismatch, $\omega_p = \sqrt{N_e e^2/\varepsilon_0 m_e}$ is the plasma frequency and $\alpha^2 = \sigma\,(c\varepsilon_0/2)^{2m} A^{2m-1} N_0 m\hbar\omega$ is the absorption due to multiphoton/tunnel ionization. We now express the complex nature of the fields in the following equation:

$$A_{\omega,3\omega} = \left|A_{\omega,3\omega}\right| \exp\left(i\phi_{\omega,3\omega}\right) \tag{6.14}$$

where $\left|A_{\omega,3\omega}\right|$ are the amplitudes $|A_\omega|$ and $|A_{3\omega}|$ and $\phi_{\omega,3\omega}$, the phases of the electric fields of the fundamental ϕ_ω and of the TH $\phi_{3\omega}$, respectively. For simplicity, we assume that there is no linear group velocity dispersion and the energy loss due to multiphoton/tunnel ionization is negligible. Substituting Eq. (6.14) into (6.12) and (6.13), separating the real and imaginary parts, one obtains the following four equations:

$$2k_{0,\omega}\frac{\partial \phi_\omega}{\partial z} = -\left(\nabla_\perp \phi_\omega\right)^2 + \frac{\Delta_\perp |A_\omega|}{|A_\omega|} - \frac{\omega_p^2}{c^2} + k_{0,\omega}^2 \frac{3}{4}\chi^{(3)} \left[\left|A_\omega\right|^2 + 2\left|A_{3\omega}\right|^2 + |A_\omega|\,|A_{3\omega}| \cos\left(3\phi_\omega - \phi_{3\omega}\right)\right] \tag{6.15a}$$

$$2k_{0,\omega}|A_\omega|\frac{\partial |A_\omega|}{\partial z} = -\nabla_\perp\left(\left|A_\omega\right|^2 \nabla_\perp \phi_\omega\right) - k_{0,\omega}^2 \frac{3}{4}\chi^{(3)} \left[|A_\omega|^3\,|A_{3\omega}| \sin\left(3\phi_\omega - \phi_{3\omega}\right)\right] \tag{6.15b}$$

$$2k_{0,3\omega}\frac{\partial \phi_{3\omega}}{\partial z} = -\left(\nabla_\perp \phi_{3\omega}\right)^2 + \frac{\Delta_\perp |A_{3\omega}|}{|A_{3\omega}|} - \frac{\omega_p^2}{c^2} + k_{0,3\omega}^2 \frac{3}{4}\chi^{(3)} \left[\left|A_{3\omega}\right|^2 + 2\left|A_\omega\right|^2 + \frac{|A_\omega|^3}{3\,|A_{3\omega}|} \cos\left(3\phi_\omega - \phi_{3\omega}\right)\right] \tag{6.16a}$$

$$2k_{0,3\omega}|A_{3\omega}|\frac{\partial |A_{3\omega}|}{\partial z} = -\nabla_\perp\left(\left|A_{3\omega}\right|^2 \nabla_\perp \phi_{3\omega}\right) - k_{0,3\omega}^2 \frac{3}{4}\chi^{(3)} \left[|A_\omega|^3\,|A_{3\omega}| \sin\left(3\phi_\omega - \phi_{3\omega}\right)\right] \tag{6.16b}$$

We observe that intensity clamping should be valid in the filament core; i.e., the fundamental's intensity does not vary in the z-direction, or $\frac{\partial |A_\omega|^2}{\partial z} \approx 0$. Also, at all these positions, self-focusing of the fundamental by the Kerr effect is balanced by the self-de-focusing effect of the plasma generated through multiphoton/tunnel ionization; i.e., the wave front is plane in the filament core or $\nabla_\perp \phi_\omega \approx 0$. Since the TH is generated at the high intensity zone of the fundamental, at these special positions inside the filament core, we also expect that the TH behaves similarly; i.e., $\frac{\partial |A_{3\omega}|^2}{\partial z} \approx 0$ and $\nabla_\perp \phi_{3\omega} \approx 0$. Using the above approximations/limiting conditions in Eq. (6.15b), one obtains the phase matching condition

$$\sin\left(3\phi_\omega - \phi_{3\omega}\right) k_0^2 \frac{3}{4}\chi^{(3)} \left|A_\omega\right|^3 |A_{3\omega}| \approx 0 \tag{6.17}$$

Equation (6.17) leads us to conclude that

$$\sin\left(3\phi_\omega - \phi_{3\omega}\right) = 0 \tag{6.18}$$

all other terms being non-zero. Thus, the condition for phase matching is given by

$$\Delta\phi \equiv 3\phi_\omega - \phi_{3\omega} = 0 \ \text{ or } \ \pi \tag{6.19}$$

i.e., the so-called phase difference ($\Delta\phi \equiv 3\phi_\omega - \phi_{3\omega}$) is a constant, either 0 or π, inside the filament core. We now apply these conditions as well as those for intensity clamping and plane wave front at the self-focal position (see discussion after Eq. 6.16b) in Eqs. (6.15a) and (6.16a) and apply the following definition for the index of refraction of the fundamental, $\partial\phi_\omega/\partial z = k_{0,\omega}n_\omega$ and that of the TH, $\partial\phi_{3\omega}/\partial z = k_{0,3\omega}n_{3\omega}$. The following results emerge:

$$n_\omega = n_{0,\omega} + \frac{3}{8}\chi^{(3)}\left[|A_\omega|^2 + 2|A_{3\omega}|^2 \pm |A_\omega||A_{3\omega}|\right] - \frac{\omega_p^2}{2c^2k_{0,\omega}^2}$$
$$= \begin{cases} 1.000315 \text{ if } \Delta\phi = 0 \\ 1.000309 \text{ if } \Delta\phi = \pi \end{cases} \tag{6.20}$$

$$n_{3\omega} = n_{0,3\omega} + \frac{3}{8}\chi^{(3)}\left[|A_{3\omega}|^2 + 2|A_\omega|^2 \pm \frac{|A_\omega|^3}{3|A_{3\omega}|}\right] - \frac{\omega_p^2}{2k_{0,3\omega}^2c^2}$$
$$= \begin{cases} 1.000401 \text{ if } \Delta\phi = 0 \\ 1.000307 \text{ if } \Delta\phi = \pi \end{cases} \tag{6.21}$$

The numerical values in Eqs. (6.20) and (6.21) are obtained using the following numerical values: $|A_\omega| \cong 20 \times 10^9$ V/m (corresponding to the clamped intensity of 5×10^{13} W/cm^2 in air), $|A_{3\omega}| \approx 3\times10^9$ V/m (corresponding to the measured intensity of 10^{12} W/cm^2, see later), the tabulated linear indices $n_{0,w} = 1.000290$ and $n_{0,3w} = 1.000311$, $\chi^{(3)}(3\omega) \approx \chi^{(3)}(\omega) = 1.4 \times 10^{-25}$ m^2/V^2 (a reasonable assumption) and the measured value of $N_e \sim 10^{16}$ electrons/cm^3 (Théberge et al., 2006a). From the numerical values in Eqs. (6.20) and (6.21), we see that the condition $\Delta\phi = \pi$ leads to an almost perfect phase matching; i.e., the indices of refraction of the fundamental and TH inside the filament core are almost identical. Hence, their speeds of propagation given by *c*/*n* where *n* is the index of refraction are almost equal during their co-propagation inside the filament core. This is self-group phase locking. The physics of this new phenomenon is evident if we take a closer look at Eqs. (6.20) and (6.21).

We shall neglect the plasma term which is weak and consider only the nonlinear terms in the square brackets

$$n_\omega - n_{0,\omega} = \frac{3}{8}\chi^{(3)}\left[|A_\omega|^2 + 2|A_{3\omega}|^2 - |A_\omega||A_{3\omega}|\right] \tag{6.22}$$

$$n_{3\omega} - n_{0,3\omega} = \frac{3}{8}\chi^{(3)}\left[|A_{3\omega}|^2 + 2|A_\omega|^2 - \frac{|A_\omega|^3}{3|A_{3\omega}|}\right] \tag{6.23}$$

Equation (6.22) shows that the index of refraction of the fundamental is dominated by the clamped intensity $|A_\omega|^2$ of the fundamental since the TH field is much weaker than that of the fundamental. This intensity term comes from self-phase modulation of the fundamental (Eq. 6.6). Meanwhile, the nonlinear index of refraction of the TH, Eq. (6.23), is also dominated by the clamped intensity of the fundamental $2|A_\omega|^2$. But this term arises from the cross-phase-modulation (XPM), Eq. (6.9), between the fundamental and the TH. The third term in Eq. (6.23) comes from THG, is also dominated by the intensity of the fundamental, and plays also an essential role in regulating the nonlinear index of the TH. The consequence is that the TH sees a nonlinear index of refraction controlled by the fundamental intensity inside the filament core. This is as if, as the fundamental self-focuses, it generates the TH and "guides" the latter to propagate faithfully with itself.

If we further look at the phase matching condition that leads the above conclusion, namely,

$$\Delta\phi \equiv 3\phi_\omega - \phi_{3\omega} = \pi \tag{6.24}$$

we realize that this is not the usual phase difference between ϕ_ω and $\phi_{3\omega}$. The above condition is obtained at the self-focal plane where there is intensity clamping and the wave fronts are plane. When the fundamental arrives at the first self-focal plane, the TH is generated "instantaneously". We can thus use the initial self-focal plane as the origin of the propagation axis. The phase accumulated after a distance of propagation inside the filament core is kz. For the fundamental, it is $\phi_\omega = k_\omega z_\omega$ and for the TH, it is $\phi_{3\omega} = k_{3\omega} z_{3\omega}$. If we substitute these two relations into Eq. (6.24), we obtain

$$\begin{aligned} 3\phi_\omega - \phi_{3\omega} &= 3k_\omega z_\omega - k_{3\omega} z_{3\omega} = 3\left(\tfrac{\omega n_\omega}{c}\right) z_\omega - \left(\tfrac{3\omega n_{3\omega}}{c}\right) z_{3\omega} \\ &= 3\left(\tfrac{\omega n_\omega}{c}\right)(z_\omega - z_{3\omega}) \quad (\because n_\omega = n_{3\omega}) \end{aligned}$$

Thus, Eq. (6.24) becomes

$$z_\omega - z_{3\omega} = \frac{\pi c}{3\omega n_\omega} \cong \lambda_\omega/6 = \lambda_{3\omega}/2 \tag{6.25}$$

The last two equalities assume that $n_\omega \approx 1$ in air. Equation (6.25) shows that after the TH is launched at the self-focal plane in air, the distance it travels along the propagation axis is consistently less than that travelled by the fundamental by a distance of about half the TH wavelength. But they both travel at the same speed (same indices of refraction). In other words, the TH is consistently lagging behind the fundamental by a distance of about half of the TH wavelength or 1/6th of the fundamental wavelength.

6.3 Experiment on THG in Air

Figure 6.1 shows the experimental set up to detect the third harmonic generation in the propagation direction (Aközbek et al., 2002a). A Ti-sapphire laser pulse at 810 nm/45 fs/2 mJ per pulse/1 kHz repetition rate was focused in air by a 100 cm lens. A 10 cm filament was generated whose diameter was about 100 μm in diameter. A diamond pinhole, 400 μm in diameter, was moved along the filament zone. The transmitted TH signal was collimated and then dispersed by two gratings and detected by a PMT. The photodiode monitored the pump signal. It should be noted that after the pinhole, the filament was terminated because the reservoir was blocked (see Section 2.3.5). Figure 6.2 shows the calibrated conversion efficiency of the TH signal as a function of the pinhole position. The transmitted pump signal of the

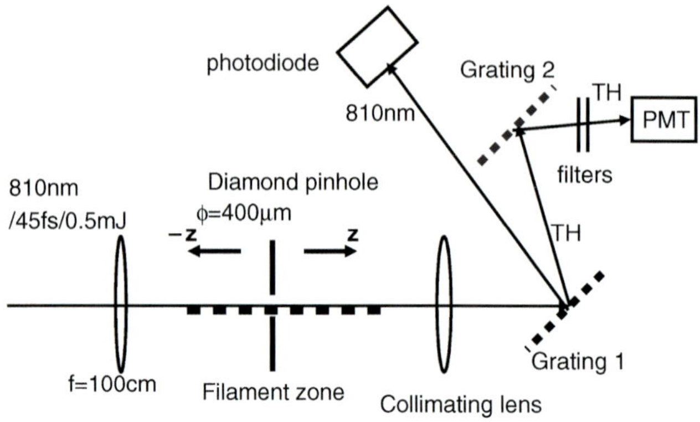

Fig. 6.1 Schematic experimental set up to detect the third harmonic (TH) generation in the propagation direction. A diamond pinhole, 400 μm in diameter, was moved along the filament zone. The transmitted TH signal was collimated and then dispersed by two gratings and detected by a PMT. The photodiode monitors the pump signal

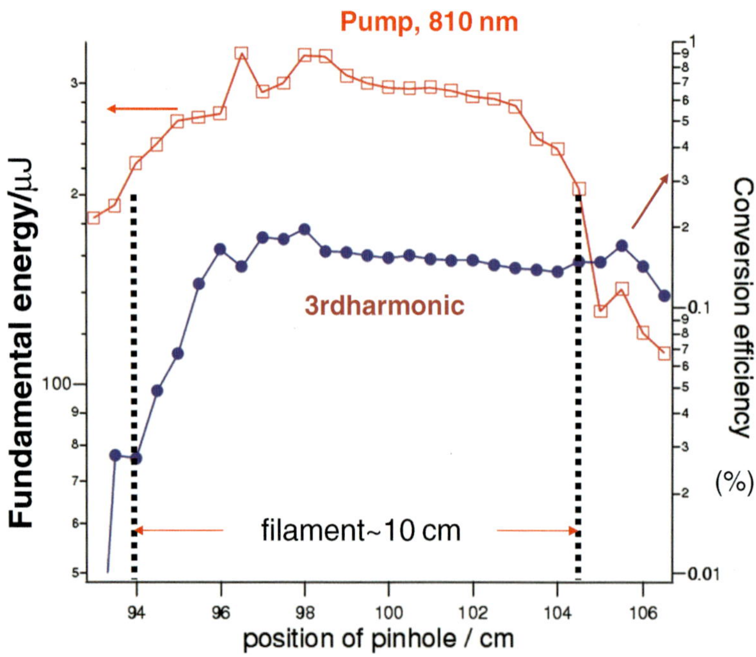

Fig. 6.2 Calibrated conversion efficiency of the TH signal as a function of the pinhole position. The transmitted pump signal of the Ti-sapphire laser pulse (810 nm/45 fs/2 mJ per pulse/1 kHz repetition rate) is also shown for comparison

Ti-sapphire laser pulse is also shown for comparison. We see that since the pinhole essentially samples the filament core where intensity clamping takes place, the transmitted pump energy is roughly constant. Consequently, the TH energy is also roughly constant over the filament zone of about 10 cm. What is worth pointing out is that the conversion efficiency is very high, of the order of 0.2%. This high efficiency indicates a good phase matching due to self-group phase locking. We note that if there were no self-group-phase locking, the conversion efficiency is estimated to be at least two orders of magnitude lower (Aközbek et al., 2002a). Furthermore, the observed stable TH energy conversion efficiency is due to intensity clamping of the fundamental in the filament core. This stabilization of both the fundamental and the TH signals is manifested in another experiment (Théberge et al., 2007a) in which a significant reduction of the root-mean-square fluctuation of the fundamental and the TH intensities by a few hundred per cent was measured experimentally.

6.4 Conical Emission and Superbroadening of the Third Harmonic in Air

As the pump power increases, conical emission of the TH occurs together with the central spot. The energy content in the conical emission of the TH is even higher than the energy in the central spot. The physical reason involves the cross phase modulation by the pump as well as the interaction with the plasma in the filament core. Detailed discussion and observation of this phenomenon is given in Théberge et al. (2005a, 2007a). This conical emission is also observed in the form of x-waves (Xu et al., 2008).

Meanwhile, as the pump power continues to increase, the TH spectral width becomes broader and broader, and soon it merges into the short wavelength side of the white light laser pulse (supercontinuum) through cross phase modulation as demonstrated by Théberge et al. (2005b). The self-broadening of the TH could be explained as being due to the self-compression of the fundamental pulse down to few cycles or less. This in turn compresses the TH pulse in the filament core significantly. This pulse shortening will naturally lead to spectral broadening.

Details of these two cases will not be discussed here and the readers are referred to the above cited references.

6.5 Efficient Tunable Few Cycle Visible Pulse Generation Through Four-Wave-Mixing Inside the Filament Core

The physics of self-group-phase locking during third harmonic generation shows that any other parametric processes should undergo similar phase locking process since THG is just a special example of the parametric process involving $\chi^{(3)}$ in a medium with inversion symmetry as predicted in Chin et al. (2005). Four-wave-mixing of two 800 nm photons (ω_{800}) from a Ti-sapphire laser and one infrared photon ($\omega_{\mathrm{i.r.}}$) tunable between 1,100 and 2,400 nm from an optical parametric amplifier ($\omega_{4\mathrm{WM}} = 2\omega_{800} - \omega_{\mathrm{i.r.}}$), is a beautiful test case since the resultant signal photon ($\omega_{4\mathrm{WM}}$) is in the visible spectral range which is easy to measure.

This experiment was carried out with great success (Théberge et al., 2006a). The experimental set up is shown schematically in Fig. 6.3. The Ti-sapphire laser pulse (CPA laser) is split into two equal parts of 1 mJ each. The fundamental at 800 nm (dashed line), after passing through a delay line, generates a filament in front of the focal region of a concave mirror. The second part of the Ti-sapphire laser pulse pumps the OPA system. The output of the OPA system is tunable from 1,100 to 2,400 nm (dotted line). It is collinear with the fundamental and is focused into the filament zone of the fundamental. The new pulse generated through four-wave-mixing (4WM) emerges collinearly (solid line). In the figure, for clarity, the three lines representing the three different pulses are separated. A prism compressor recompresses

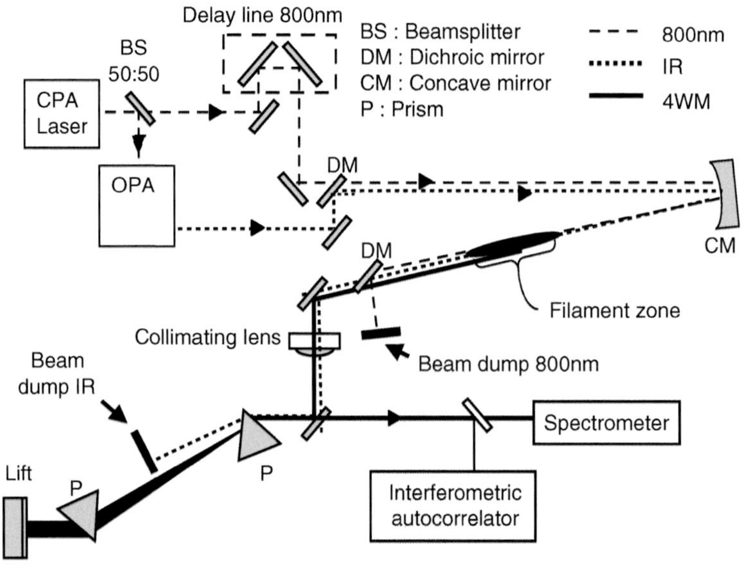

Fig. 6.3 Four wave mixing inside a filament. The Ti-sapphire laser pulse (CPA laser) is split into two equal parts of 1 mJ each. The fundamental at 800 nm (*dashed line*), after passing through a delay line, generates a filament in front of the focal region of a concave mirror. The second part of the Ti-sapphire laser pulse pumps the OPA system. The output of the OPA system is tunable from 1100 nm to 2400 nm (*dotted line*). It is collinear with the fundamental and is focused into the filament zone of the fundamental. The new pulse generated through four wave mixing (4WM) emerges collinearly (*solid line*). In the figure, for clarity, the three lines representing the three different pulses are separated. A prism compressor recompresses the 4WM pulse which is then measured by an interferometric autocorrelator and a spectrometer. The wave length of the 4WM pulse is tunable across almost the entire visible range

the 4WM pulse which is then measured by an interferometric autocorrelator and a spectrometer.

Efficient generation of the 4WM pulse is obtained when the 800 nm and the i.r. pulses overlap with each other. The wavelength of the 4WM pulse is tunable across almost the entire visible range. Its duration depends on the length of the filament. The longer the length of the filament is, the shorter is the 4WM pulse. This is reasonable since the longer the filament is, the shorter the 800 nm pulse is self-compressed to. Since the i.r. pulse is phase locked with the 800 nm pulse (i.e., they propagate and interact together at the same speed), its duration naturally follows that of the 800 nm pulse.

Self-group-phase locking of the i.r. pulse by the 800 nm filamenting pulse is manifested by two observations of the resultant 4WM pulse. The first is that the maximum conversion efficiency of the 4WM pulse (defined as the ratio of the energy of the 4WM pulse over that of the tunable i.r. pulse) is less than 10^{-5} when the peak power of the 800 nm pulse is below the critical power for self-focusing P_c in air (i.e., no filament is created). It jumps to 30% (in air) once the peak power of the 800 nm pulse is higher than P_c when a filament is formed. In argon gas at 1.5 atmospheric pressure, it can reach up to 60%. Such high efficiencies could not have been reached without phase matching.

The other manifestation is the stability of the 4WM pulse. The root-mean-square (RMS) fluctuation of the 800 nm pulse and that of the i.r. pulse are 1.5 and 1.6% respectively. Without filamentation, the generated 4WM pulse has a RMS fluctuation of 5.2%. This follows from statistical considerations. The expected RMS fluctuation of the 4WM pulse should be approximately the sum of twice that of the 800 nm pulse plus that of the i.r. pulse ($\mathrm{RMS}_{4\mathrm{WM}} \cong 2\mathrm{RMS}_{800} + \mathrm{RMS}_{\mathrm{i.r.}}$; RMS means the RMS fluctuation). This sum is 4.8%, which agrees with the measured value rather well. But when the peak power of the 800 nm pulse is higher than P_c, the RMS fluctuation of the 4WM pulse reduces to 1.8%. This is possible because the 4WM pulse is locked to the filamenting pulse at the filament core. The real fluctuation in the filament core is much smaller than that measured for the whole 800 nm pulse. In fact, because of intensity clamping inside the filament core, we should expect very little fluctuation inside the core. However, there is no easy and direct means to take the measurement inside the core. 4WM becomes a probe when the peak power of the filamenting pulse is higher than P_c. This is because now the 4WM pulse is phase locked with the filamenting pulse in the core only. This phase locking is manifested in the measured fluctuation. If we again use the relation

$$\mathrm{RMS}_{4\mathrm{WM}} \cong 2\mathrm{RMS}_{800} + \mathrm{RMS}_{\mathrm{i.r.}} \tag{6.26}$$

the RMS fluctuation of the 800 nm filament core is in fact unknown while the others are already measured (1.8 and 1.6% for the 4WM pulse and the i.r. pulse, respectively). Solving this equation yields $\mathrm{RMS}_{800} \cong 0.1\%$. That is to say, the measured fluctuation of the 4WM pulse reflects only the fluctuation of the i.r. pulse which does not undergo self-focusing. The influence of the very stable filamenting pulse is negligible. One could imagine that if the i.r. pulse were strong enough to generate its own filament, the resulting 4WM pulse would be as stable as the two filamenting cores. This idea remains to be tested experimentally.

Self-spatial filtering of the filamenting pulse is also manifested beautifully in this work. Since the 4WM pulse samples the filament core, its transverse fluence distribution should reflect that of the spatially-filtered filament core. Since only the fundamental mode of the 800 nm pulse undergoes filamentation, the beam quality of the 4WM pulse should reflect this nature and should thus be almost perfect. The experiment uses two 800 nm filamenting pulses of very different quality. One is the normal output pulse of the Ti-sapphire laser. The other is the 800 nm pulse diffracted through a square opening giving rise to some "ugly" diffraction pattern. Both pulses yield the same beam quality for the 4WM pulse with a $M^2 \leq 1.01$. This kind of quality can very seldom be obtained in any other normal laser system.

6.6 Self-Group-Phase Locking During Four-Wave-Mixing Inside a Filament

There is not yet a theory that simulates the above experimental results. However, our understanding of the THG could guide us to appreciate the self-group-phase locking process in all 4WM processes inside a filament. Let us first look at the current 4WM results. Phase matching of the 800 nm filamenting pulse and the i.r. pulse leads to the efficient generation of the visible 4WM pulse which itself should also be phase locked to the 800 nm pulse. We can understand the last statement from the consideration in THG. The TH pulse is locked to the 800 nm pulse due to cross-phase-modulation (XPM) indicated by Eqs. (6.9) and (6.23). Analogically, we could express the material polarization at the i.r. frequency due to XPM as

$$P^{(3)}_{\text{i.r.}} \propto |A_\omega|^2 A_{\text{i.r.}} e^{i(k_{ir}z-\omega_{ir}t)} \tag{6.27}$$

The corresponding term due to XPM between the 4WM and the i.r. pulses $P^{(3)}_{\text{i.r.}} \propto |A_{4\text{WM}}|^2 A_{\text{i.r.}} e^{i(k_{\text{i.r.}}z-\omega_{\text{i.r.}}t)}$ is neglected since this intensity of the 4WM pulse is much lower than that of clamped intensity of the 800 nm pulse (Eq. 6.27).

The material polarization at the i.r. frequency also possesses a term due to 4WM given below:

$$P^{(3)}_{\text{i.r.}} \propto |A_\omega|^2 A_{4\text{WM}} e^{i(k_{\text{i.r.}}z-\omega_{\text{i.r.}}t)} \tag{6.28}$$

Both terms in Eqs. (6.27) and (6.28) show the control of the material polarization, hence, the propagation speed (see Eq. 6.11) by the clamped intensity of the filament. The 4WM pulse is also similarly controlled by the clamped intensity of the filamenting pulse through analogous considerations of the material polarization at the 4WM frequency

$$P^{(3)}_{4\text{WM}} \propto |A_\omega|^2 A_{\text{i.r.}} e^{i(k_{4\text{WM}}z-\omega_{4\text{WM}}t)} \tag{6.29}$$

and

$$P^{(3)}_{4\mathrm{WM}} \propto |A_\omega|^2 A_{4\mathrm{WM}} e^{i(k_{4\mathrm{WM}} z - \omega_{4\mathrm{WM}} t)} \tag{6.30}$$

The influence of the i.r. intensity is neglected. Because of the above mentioned control of the indices of refraction of the i.r. and the 4WM pulses by the 800 nm filamenting pulse's high intensity inside the filament core, it is expected that they all would travel together at the same speed similar to the case of the THG. In fact, the experimental results on 4WM in the last section (Section 6.5) indicates that such phase locking does exist, which results in high conversion efficiency and high beam quality as well as self-stabilized operation.

We can extrapolate the above consideration to any 4WM process inside a filament. One would expect that with the appropriate combination of frequencies, one should expect to generate pulses of any wavelength of the electromagnetic wave from the radio frequency (even DC field) to the ultraviolet and beyond. So long as one pulse is filamenting, similar results should be obtained as in the preceding section.

For example, tera-hertz pulses in air have been generated by mixing the second harmonic of an 800 nm fs laser pulse inside the filament of another 800 nm pulse (Xie et al., 2006). Deep u.v. pulses were also generated in gases through similar 4WM scheme (Fiji et al., 2007).

This is still a young field of research. A lot of new results, phenomena and applications will certainly emerge in the course of time now that filamentation has attracted a lot of attention in the field of ultrafast intense laser science.

6.7 Derivation of Equation (6.1)

Before ending, a brief explanation of the derivation of Eq. (6.1) is given. For detail, see Aközbek et al. (2002a, 2003) and Théberge (2007).

Starting from the Maxwell's equations, one will obtain the nonlinear wave equation

$$\nabla^2 \vec{E} - \mu_0 \frac{\partial^2 \vec{D}^{(1)}}{\partial t^2} = \mu_0 \frac{\partial^2 \vec{P}^{NL}}{\partial t^2} \tag{6.31}$$

where $\vec{E}$ denotes the electric field; $\vec{D}^{(1)}$ comes from $\vec{D}$ which represents the electric flux density, and μ_0 is the magnetic permeability. $\vec{D} = \varepsilon_0 \vec{E} + \vec{P}^{(1)} + \vec{P}^{NL}$; $\vec{D}^{(1)} = \varepsilon_0 \vec{E} + \vec{P}^{(1)}$; $\vec{P}^{(1)} = \varepsilon_0 \chi^{(1)} \vec{E}$; $\vec{P}^{(NL)} = \varepsilon_0 (\chi^{(2)} \vec{E} + \chi^{(3)} \vec{E}\vec{E} + ...)\vec{E}$. $\chi^{(1)}$ is the linear susceptibility and $\chi^{(2)}$, $\chi^{(3)}$,... are the nonlinear susceptibility of the medium. The wave equation is valid for all frequency components that constitute the ultrafast laser pulse. It is thus necessary to decompose the electric field into its frequency components through Fourier transform and rewrite Eq. (6.31) in the frequency domain. We can write the electric field propagating in the z-direction as $E(r,z,t) = A(r,z,t)\, e^{-i(\omega_0 t - k_0 z)} + c.c.$ where $c.c.$ is the complex conjugate of the field; A is the complex envelope; ω_0, the central frequency of the pulse; and k_0 is the linear wave vector. We now express the vectors $\vec{V} \equiv \vec{E},\vec{D},\vec{P}^{(NL)}$ as the Fourier transform

$$\begin{aligned}\vec{V}(r,z,t) &= \frac{1}{2\pi}\int_0^{+\infty} \tilde{V}(r,z,\omega)\,e^{i\omega t}d\omega + c.c.\\ \tilde{V}(r,z,\omega) &\equiv \tilde{E}(r,z,\omega),\,\tilde{D}(r,z,\omega),\,\tilde{P}(r,z,\omega)\end{aligned} \tag{6.32}$$

The susceptibility of the medium becomes a tensor ($\chi^{(1)}$, $\chi^{(2)}$, $\chi^{(3)}$,...) that has a relationship with the complex fields given below:

$$\tilde{D}^{(1)}(r,z,\omega) = \varepsilon_0\tilde{E}(r,z,\omega) + \varepsilon_0\chi^{(1)}(\omega)\cdot\tilde{E}(r,z,\omega) \tag{6.33}$$

$$\tilde{P}^{NL}(r,z,\omega) = \varepsilon_0\left(\chi^{(2)}\cdot\tilde{E}(r,z,\omega) + \chi^{(3)}\cdot\tilde{E}^2(r,z,\omega) + \cdots\right)\cdot\tilde{E}(r,z,\omega) \tag{6.34}$$

Using these relations (6.32), (6.33), and (6.34) in Eq. (6.31), one obtains

$$\nabla^2\tilde{E} + \frac{\omega^2}{c^2}\left(1+\chi^{(1)}\right)\cdot\tilde{E} = -\mu_0\omega^2\tilde{P}^{NL} \tag{6.35}$$

We use the following relationships:

$$\chi^{(1)} = \chi^{(1)'} - \left(\omega_p^2/\omega^2\right) + i\chi^{(1)''} \tag{6.36}$$

where the linear susceptibility $\chi^{(1)}$ is expressed in terms of the imaginary part $\chi^{(1)''}$ and the real parts (loss due to material's weak absorption $\chi^{(1)'}$ and plasma generation through the plasma frequency $\omega_p = \sqrt{N_e e^2/\varepsilon_0 m_e}$ where N_e is the electron density, m_e and e are the mass and the charge of the electron). The linear wave vector is given by $k = \frac{\omega}{c}\sqrt{1+\chi^{(1)'}}$ and the absorption coefficient, $\alpha = (\omega/c)\sqrt{\chi^{(1)''}}$. Expressing $\nabla^2 = \Delta_\perp + \frac{\partial^2}{\partial z^2}$, Eq. (6.35) becomes

$$\Delta_\perp\tilde{E} + \frac{\partial^2\tilde{E}}{\partial z^2} + k^2\cdot\tilde{E} - \frac{\omega_p^2}{c^2}\tilde{E} + i\alpha^2\tilde{E} = -\frac{k^2}{\varepsilon_0}\tilde{P}^{NL} \tag{6.37}$$

A solution of this equation is the travelling wave $\tilde{E}(r,z,\omega) = \tilde{A}(r,z,\omega)\,e^{ik_0z} + c.c.$ where k_0 is the linear wave vector for the central frequency ω_0. We put the travelling wave solution into Eq. (3.37), dividing by e^{ik_0z} and use the approximation that $k \approx k_0$, so that $\left(k^2 - k_0^2\right) = (k + k_0)(k - k_0) \cong 2k_0(k - k_0)$. Also used is the dispersion relation of k

$$k = k_0 + (\omega - \omega_0)\, k' + \frac{1}{2}(\omega - \omega_0)^2\, k'' + \cdots \tag{6.38}$$

$$k' = \left.\frac{\partial k}{\partial \omega}\right|_{\omega_0} = \frac{1}{v_g}$$

$$k'' = \left.\left(\frac{\partial^2 k}{\partial \omega^2}\right)\right|_{\omega_0} = \left.\left(\frac{-1}{v_g}\frac{\partial v_g}{\partial \omega}\right)\right|_{\omega_0}$$

where v_g is the group velocity; k'' is group velocity dispersion (GVD) parameter.

Performing the inverse Fourier transform of the resulting equation back to the time domain, one obtains

$$\Delta_\perp A + \frac{\partial^2 A}{\partial z^2} + 2ik_0\left[\frac{\partial A}{\partial z} + \frac{1}{v_g}\frac{\partial A}{\partial t}\right] - k_0 k''\frac{\partial^2 A}{\partial t^2} - \frac{\omega_p^2}{c^2}A + i\alpha^2 A = -\frac{k_0^2}{\varepsilon_0}P^{NL}e^{-ik_0 z} \tag{6.39}$$

where

$$A\,(r,z,t) = \frac{1}{2\pi}\int_0^{+\infty} \tilde{A}\,(r,z,\omega)\, e^{i\omega t} d\omega + c.c. \tag{6.40}$$

Using the slowly varying envelope approximation (SVEA)

$$\frac{\partial A}{\partial z} \ll k_0 A; \quad \frac{\partial^2 A}{\partial z^2} \ll k_0 \frac{\partial A}{\partial z} \tag{6.41}$$

One obtains the final equation (6.1)

$$\Delta_\perp A + 2ik_0\left[\frac{\partial A}{\partial z} + \frac{1}{v_g}\frac{\partial A}{\partial t}\right] - k_0 k''\frac{\partial^2 A}{\partial t^2} - \frac{\omega_p^2}{c^2}A + i\alpha^2 A = -\frac{k_0^2}{\varepsilon_0}P^{NL}e^{-ik_0 z} \tag{6.42}$$

Chapter 7
Remote Sensing Using Filamentation

7.1 Introduction

Laser-based spectroscopic techniques, such as differential absorption LIDAR (Light Detection And Ranging), tunable diode laser absorption spectroscopy, and laser-induced fluorescence have been extensively employed for sensing atmospheric trace species because of their high sensitivity, non-intrusiveness, and real-time analysis (Fehér and Martin, 1995; Svanberg, 2004). However, with these techniques, the laser usually may only be optimized one pollutant at a time.

The Teramobile group has made use of the back-scattering of the white light laser pulse propagating in air as a "lamp in the sky". Because of the continuous broadband spectrum of the white light laser pulse spanning from the i.r. to the near UV, many pollutant molecules in air can absorb the back-scattered light. They measured the linear absorption spectra of some molecules such as water (humidity) and ozone from long distances in air, showing the feasibility of this technique (Kasparian et al., 2003).

The author proposes a new LIDAR technique based on filamentation for the detection of chemical and biological agents or pollutants. Such filamentation can be controlled, in principle, to occur at a distance as far as a few kilometers in the atmosphere (Chin et al., 2005; Rodriguez et al, 2004). Inside the filament, the peak intensity is clamped around 5×10^{13} W/cm^2 (Chin et al., 2005) and is high enough to dissociate/ionize gas molecules, to explode fine particles (dusts and aerosols) and to induce "rapid or partial" breakdown on solid targets. It was predicted by the author (Chin, 2007) that the interaction of the strong field inside the filament in air with molecules would result in "clean" fluorescence emissions practically free of plasma continuum (Talebpour et al., 2000). We further discovered that such "clean" fluorescence is a general phenomenon in the interaction of the ultrafast strong field inside the filament and all targets tested so far, ranging from gases, vapors, aerosols to solids. Time-resolved technique could be used to bring out such "clean" fluorescence (Luo et al., 2004). Furthermore, such "clean" fluorescence is unique from agent to agent; i.e., each agent will result in its own finger print fluorescence (Gravel et al., 2004). We have carried out systematic experimental studies inside the laboratory environment by measuring the characteristic finger print fluorescence of

S.L. Chin, *Femtosecond Laser Filamentation*, Springer Series on Atomic, Optical, and Plasma Physics 55, DOI 10.1007/978-1-4419-0688-5_7,

representative target examples. These include gases (CH_4 (Xu et al., 2006a); C_2H_2 (Xu et al., 2007c)), vapor (ethanol (Luo et al., 2006)), bio-agents (powders of egg white and yeast (Xu et al., 2006b), grain dusts of barley, corn and wheat (Xu et al., 2007b)), water aerosols containing multiple solutes: NaCl, $PbCl_2$, $CuCl_2$, and $FeCl_2$ (Daigle et al., 2007a, b) and metallic targets (lead, copper and aluminum) (Liu et al., 2007; Xu et al., 2007a). The above mentioned laboratory results can be extrapolated to remote targets using the LIDAR equation (Measures, 1992). A simplified form is given below:

$$S = C\frac{\exp^{-\alpha r}}{r^2} \tag{7.1}$$

where S is the detected signal, r is the distance between the detector and the target, α is the absorption coefficient in the detection path, and C is a parameter that depends on the atmospheric optical condition, the target fluorescence's characteristics, the detection geometry, and the detector parameters. Assuming the same operating conditions as in the laboratory studies, the extrapolation depends only on the distance r. We show that it is feasible to remotely detect all these agents in the range of kilometers. This opens the door toward remote detection of targets related to safety, security, and pollution. In particular, one laser (in our case, the femtosecond Ti-sapphire laser) could do all the detections so far. A detailed review of this new technique can be found in Chin et al. (2009).

7.2 Remote Control of Filamentation

In remote detection of a gaseous target using the LIDAR technique, it is desirable to be able not only to identify the target but also to locate its position. Thus a very long filament would not be suitable, while too short a filament might not produce enough signals. We have developed a simple technique that satisfies both requirements, at least in the laboratory environment.

In Chapter 4, we discussed multiple filaments competition. We have measured the back-scattered nitrogen fluorescence from a distance. If the diameter of the beam is large (25 mm), each hot zone in the transverse plane of the pulse will compete for energy in the reservoir, resulting in a large fluctuation of the signal (Fig. 7.1a). When the diameter of the beam is reduced (squeezed) by about three times (8 mm) using an inverted telescope while keeping the energy of the laser pulse constant, the detected back-scattered nitrogen fluorescence from the filaments becomes much more stable from shot to shot (Fig. 7.1b). Furthermore, the fluorescence signal increases by about three orders of magnitude (Luo et al., 2005). This can be interpreted, through numerical calculation (Kosareva et al., 2006), as being due to the constructive interferences of filament zones which are now near one another resulting in a faster growth of multiple filaments with propagation distance, a larger average diameter of plasma channels, and a larger overall amount of electrons in the transverse beam section.

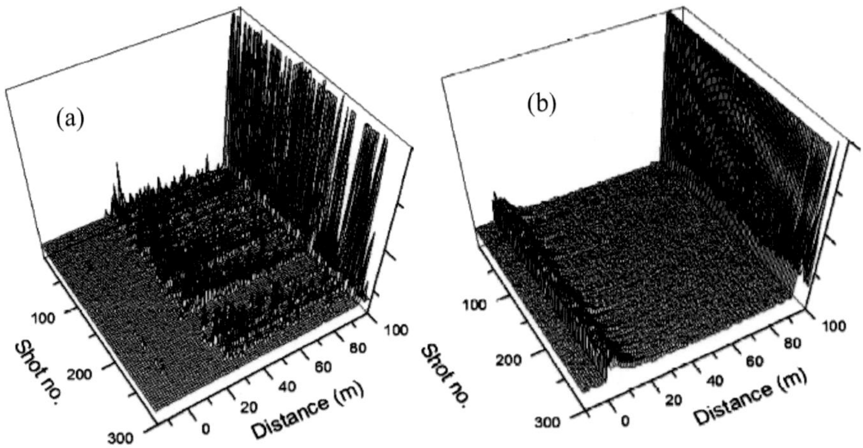

Fig. 7.1 Shot-to-shot back-scattered fluorescence waveforms from excited N_2 at 337 nm using a PMT. The laser energy was fixed at 30 mJ/pulse. The diameters of the beams are (**a**) 25 mm and (**b**) 8 mm, respectively. There are 300 single-shot traces of the fluorescence signal versus the distance in each graph

However, only using a small beam diameter does not permit us to move the high intensity filaments to a long distance. This is because the diameter of each of the hot zones becomes smaller so that its self-focusing distance becomes short. Even by changing the initial divergence and the chirp would not help moving the strong plasma zone significantly. We overcome such early self-focusing by using a telescope (Liu et al., 2006b) which enlarges the diameter of the beam, thus that of the hot spots. The telescope's effective focal length is much shorter than the self-focusing distance of both the enlarged beam and the hot spots. Under this condition, the resulting multifilaments merge into the geometrical focus, which is adjustable by varying the relative distance between the divergent and convergent optical components, as shown in Fig. 7.2. In this experimental scheme, filamentation starts near

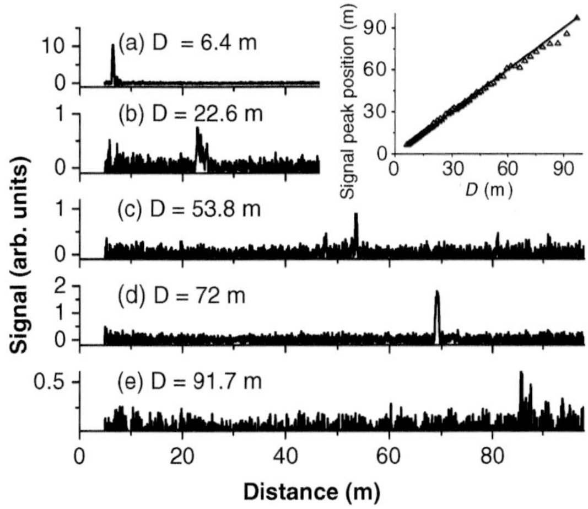

Fig. 7.2 Lidar collected 337 nm signals as function of the distance for different sending telescope configurations. D indicates the distance between the convex mirror ($f = -50$ cm) and the calculated geometrical focus position. *Inset*: comparison between the measured fluorescence peak position (*open triangles*) and the calculated geometrical focus (*solid line*: slope = 1)

the geometrical focus and the beam size is small at this position. This is equivalent to a small beam diameter and will produce strong fluorescence signals. Considering that we need to spatially resolve the fluorescence zone reasonably from a distance, this technique simultaneously solves the problem of shortening the filament length because now the strong filament zone is limited to around the Rayleigh range of the focusing telescope.

7.3 Physical Considerations

In atmospheric remote sensing, it is also of particular importance to increase the signal-to-noise ratio. In our case, one noise source is the white light (supercontinuum) produced during the filamentation process in air due to self-phase modulation and self-steepening of the laser pulses (see Chapter 2). This white light spectrum can span from the ultraviolet (UV) to the infrared (IR) and may mask the fluorescence signals from the sample. Therefore, there is always a question whether the fluorescence from molecules can be distinguished from the backscattered white light.

To answer this question, we show experimentally the capability of distinguishing the nitrogen fluorescence spectrum emitted from inside the filament from the back-scattered supercontinuum both spectrally and over a long range in atmospheric air (Théberge et al., 2005a). The spectral broadening of the laser pulse develops progressively along the self-induced plasma column in the propagation direction in air. The broad spectrum of the back-scattered white light laser pulse becomes fully developed at the end of the plasma column; this is the result of the distance-cumulative effects of self-phase modulation and self-steepening of the fundamental laser pulse. Therefore, throughout the whole filament, the strong backward-emitted nitrogen fluorescence is clearly resolved from the back-scattered white light laser

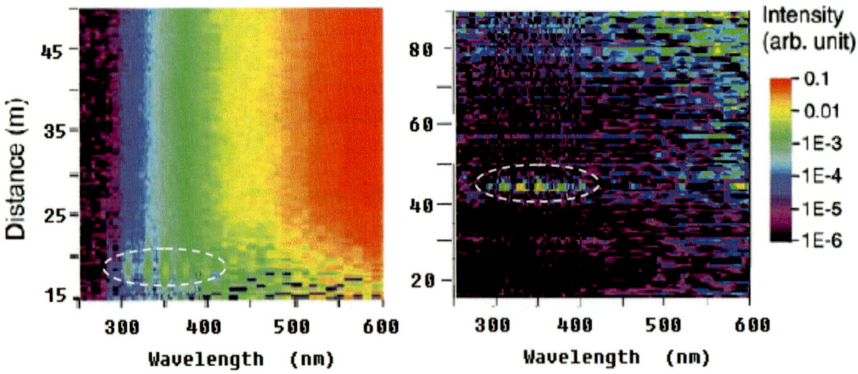

Fig. 7.3 Intensity distribution of the back-scattered signal from 250 to 600 nm as a function of the distance and the wavelength. *Left*: using a collimated laser beam; *right*: using a focused beam by the telescope with $D = 45$ m. The *white dash ellipses* highlight the resolved nitrogen fluorescence

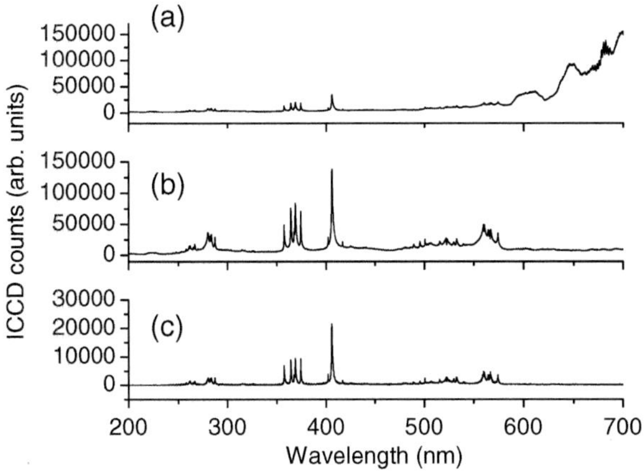

Fig. 7.4 Spectra of filament-induced lead plasma. The sample is located at (**a**, **c**) 4.8 m and (**b**) 2.8 m away from the focusing lens ($f = 5$ m). The gate widths are $t = 2$ μs (**a**, **b**, and **c**) and the delay times are $\Delta t = -3$ ns (**a** and **b**) and $\Delta t = 20$ ns (**c**), respectively. Note that $t = 0$ is the laser arriving time on the target and the filament started at a distance around 2.5 m away from the focusing lens

pulse (supercontinuum) (Fig. 7.3, left). Based upon this finding, we can minimize the white light (Liu et al., 2006b) by producing strong but short filaments at a remote focus (Fig. 7.3, right). For solid target detections, the white light can be minimized by using the beginning of the filaments to interact with the sample, or to use a time-resolved measurement (Fig. 7.4). Moreover, the characteristic fluorescence of atmospheric pollutants usually has a lifetime longer than that of nitrogen. Using electronic delay and gating techniques, we can temporally resolve this from the nitrogen fluorescence and the back-scattered white light laser pulse, therefore enhancing the signal-to-noise ratio for remote detection.

7.4 Detection of Chemical and Biological Agents in Air

Based on the technology developed so far and the understanding of the filamentation physics (Chin, 2006; Chin et al., 2005, 2007, 2008a), we performed a spectroscopic study of the interaction of the filaments with different targets including gases, vapors, solids and aerosols, in order to show the feasibility that intense femtosecond laser pulses could be applied to remote sensing of chemical/biological agents using only one laser.

7.4.1 Molecules in the Gas/Vapor Phase

The fluorescence spectrum of ethanol vapor induced by the filamentation in air has been recorded as the fingerprint of the molecule (Luo et al., 2006). Ethanol was selected because it is not harmful for human health when compared with most of the hydrocarbon molecules which are toxic and unsuitable for open-air experiments in the laboratory. This is the first observation of back-scattered fluorescence from the hydrocarbon pollutants, but with a constant concentration of 6.8%.

Next, we design a long vacuum tube (4.5 m) for sensing the greenhouse hydrocarbon gas methane (Xu et al., 2006a). The reason for designing this long vacuum tube is to be able to electronically gate out the scattered white light induced by the last window of the tube and to allow spectral measurement over a reasonable period of time after the passage of the laser pulse. Backward fluorescence from dissociated CH radicals is used to analyze quantitatively the concentration of CH_4 and its remote detection limit (Fig. 7.5). The estimation based on the experimental results shows that the concentration sensitivity could be down to the ppm range, and the detection range limit could extend up to the kilometer range. The dissociation mechanism of CH_4 has been discussed in Kong et al. (2006). Briefly, the intense laser field, at an intensity around or above 10^{14} W/cm^2, weakens the molecular chemical bonds and causes polyatomic molecules to dissociate into small neutral fragments. The tetrahedral methane molecule thus undergoes a stepwise disintegration. The hydrogen atoms are individually cleaved from the CH_4 molecule resulting in the production of excited CH radicals, which fluoresce.

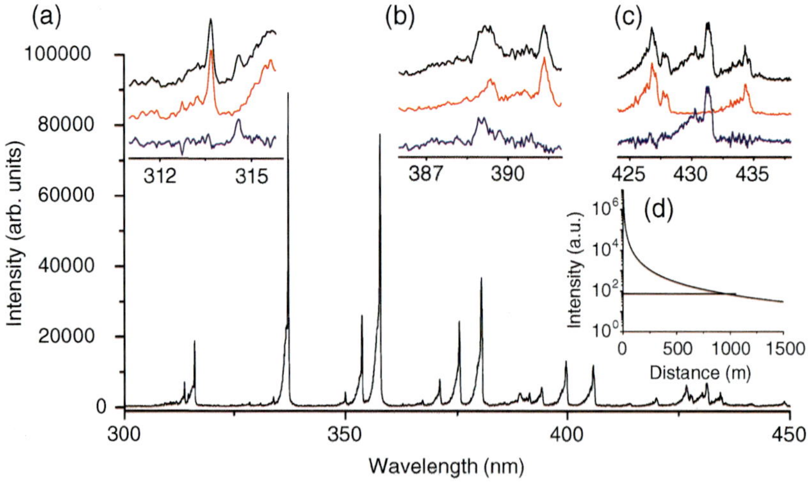

Fig.7.5 Filament-induced fluorescence spectrum of mixture of CH_4 and air with a CH_4 concentration of 2.6% (volume/volume). The insets (**a**, **b** and **c**) show the spectra in a higher resolution (*top*), the spectra of pure air in atmospheric pressure (*middle*), and the subtraction of the mixture and pure air spectra (*bottom*). The bands are (**a**) CH: C $^2\Sigma^+ - X\,^2\Pi$, (**b**) CH: B $^2\Sigma^- - X\,^2\Pi$, and (**c**) CH: A $^2\Delta - X\,^2\Pi$. The inset (**d**) shows the extrapolation of the detection limit according to the LIDAR equation ($I = L/R^2$, where I is the signal intensity, L the effective filament length and R the distance between the end of the filament and the detector). The detection limit is about 0.9 km for the CH_4 concentration of 5% and the filament length of 20 m

Next, we discuss the simultaneous detection and identification of two unknown trace gases in the atmosphere using filament-induced nonlinear spectroscopy (FINS) (Xu et al., 2007b). A genetic algorithm has been used to identify the unknown spectra with the premise that a spectral database including the spectral signatures and the strengths of the signals of the corresponding trace species is built. The detection sensitivity is in the ppm to ppb concentration range, depending on the induced fluorescence efficiency from the molecules (Fig. 7.6). The agreement between calculation and experiment observed in the specific case of methane and acetylene opens a door for a future performance of multiple species atmospheric sensing and verifies that a single laser is sufficient to induce characteristic fluorescence from a large number of molecular species.

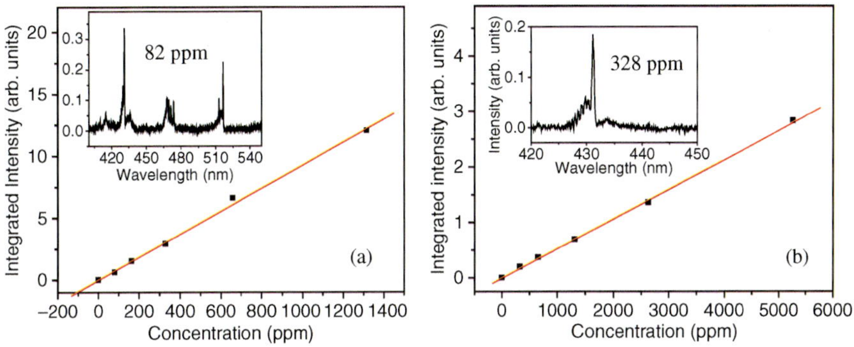

Fig. 7.6 Fluorescence signals (*rectangular points*) as function of (**a**) C_2H_2 and (**b**) CH_4 concentrations together with linear fits (*solid line*). The insets show part of the filament-induced fluorescence spectra of C_2H_2 (82 ppm) and CH_4 (328 ppm). The CH at 431 nm and the C_2 bands at 470 and 515 nm can be clearly observed. The 3σ (σ is the standard deviation of the background noise level) detection sensitivities are 350 ppb and 2 ppm for C_2H_2 and CH_4, respectively

7.4.2 Biological Targets

For the detection of solid bio-chem samples, we have experimentally shown the remarkably distinct spectra of egg white and yeast powders using time-resolved filament-induced breakdown spectroscopy (FIBS) (Xu et al., 2006b). In particular, we demonstrated the feasibility of remote detection and differentiation of some very similar agriculture related bio-aerosols, namely barley, corn, and wheat grain dusts using this technique (Xu et al., 2007b). The signals were detected in Lidar configuration. All the species showed identical spectra, namely those from molecular C_2 and CN bands, as well as atomic Si, C, Mg, Al, Na, Ca, Mn, Fe, Sr and K lines. These identical spectral bands and lines reveal similar chemical compositions; however, the relative intensities of the spectra are different showing different element abundances from these three bio-targets. The intensity ratios of different elemental lines were used to distinguish these three samples.

7.4.3 Metallic Targets

We have also demonstrated that the usage of a simple telescope as sending optics (Liu et al., 2006b) could greatly improve the performance of remote FIBS (R-FIBS) of metallic targets (lead, aluminum and copper) (Liu et al., 2007). In this case, because the filaments are short, white light continuum inside R-FIBS spectrum is negligible, realizing non-gated R-FIBS. And because the filaments are strong, the resulting line emission is impressively intense. The extrapolated detection limit of the aluminum sample reaches a few kilometers in distance and a few ppm in terms of minor element concentration when the sample is located 50 m away from the detection system.

7.4.4 Water Aerosols Containing Metallic Salts

Finally, R-FIBS was used for probing a cloud of micro-droplets where table salt had been dissolved (Daigle et al., 2007a). These micro-droplets are a good stimulant for aerosols. We demonstrated experimentally that R-FIBS can efficiently be used

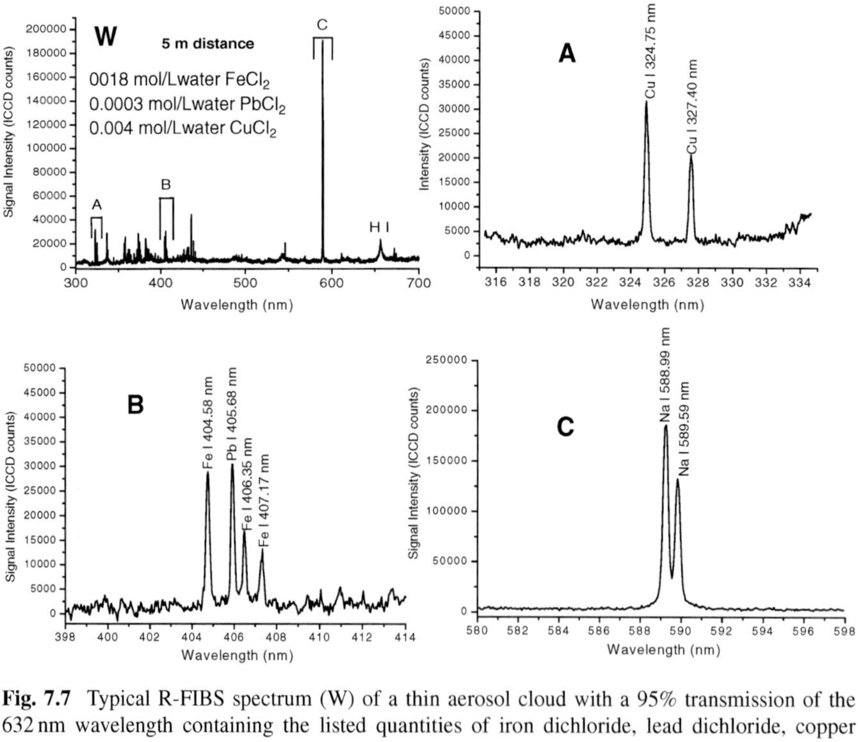

Fig. 7.7 Typical R-FIBS spectrum (W) of a thin aerosol cloud with a 95% transmission of the 632 nm wavelength containing the listed quantities of iron dichloride, lead dichloride, copper dichloride and sodium chloride. The regions referred to as A, B, and C in W are enlarged for clarity. **A** corresponds to Cu I, **B** corresponds to Fe I and Pb I and **C** corresponds to Na I

as a ppm level sensing technique remotely to retrieve the composition of micro-droplets in clouds located at a distance. The technique has been successfully tested up to 70 m and as revealed by extrapolation, showed great potential for kilometer range application. This technique is sensitive to the solvent as well. Four hydrogen bands from the Balmer series were observed in a cloud of aqueous microdroplets after H_2O molecules were broken by the light filaments. Additionally, a cloud of aqueous aerosols containing a mixture of $PbCl_2$, $CuCl_2$, $FeCl_2$ and NaCl has also been detected using R-FIBS (Daigle et al., 2007b). It was found that fluorescence from all the metallic ions dissolved can be observed (Fig. 7.7). Moreover, these spectrally narrow atomic transitions excited by the low density plasma did not show any signal overlap.

7.5 Conclusion and Looking Ahead

In this chapter, based on the experimental evidence, we conclude that "clean" fluorescence emission of bio-chem agents induced by the filaments of powerful femtosecond laser pulses in air, can indeed be used as the fingerprint to distinguish different chemical and biological agents. A single laser is sufficient to induce characteristic fluorescence from a large number of molecular species and the possibility of observing many atmospheric constituents of interest is not to be questioned anymore. This allows us to propose that femtosecond laser filamentation would be an "ideal tool" for the detection and identification of atmospheric chemical and biological agents/pollutants. However, all the experiments were performed in a laboratory scale with a limited distance of up to 100 m; remote sensing in the atmosphere lacks long distance outdoor tests and additional effort is thus needed for practical applications of this technique. Furthermore, remote projection of multiple ionizing filaments (see Chapter 4) is still a challenge for practitioners in this direction, mainly because of bad beam quality that could be affected not only intrinsically by the current laser technology but also by the medium inhomogeneity (pressure gradient, temperature gradient, etc.).

The use of deformable mirrors to improve the beam quality has just started in some laboratories, including our laboratory. By correcting only partially the astigmatism of the telescopic system used in the above experiments, we have seen an improvement of more than an order of magnitude of the return nitrogen fluorescence signal from a distance of up to 90 m (Daigle et al., 2008) inside the laboratory.

The future challenge is not whether the idea of the author would work. It does work; i.e., the strong field generated inside the filament indeed can fragment molecules and any target resulting in the emission of their characteristic finger print fluorescence. The real challenge is both technical and cost effectiveness. On the technical side, one needs to be able to control the position of the strong filamentation zone to long distances of up to 1–2 km. But even a few hundred meters is already good for certain environmental applications. The other obstacle is the cost of the laser system which is prohibitively high and TW level laser is not yet a turn-key stable system immune to vibration and temperature fluctuation in the field.

Chapter 8
Challenges Ahead

There are still many challenging problems to be solved. From the point of view of propagation, multiple filaments and white light control are among these challenges. From the interaction physics point of view, a new direction in strong field molecular physics is being opened up. The latter is the consequence of the observation of clean fluorescence from molecules inside the filaments in air (Chapter 7). In the beginning of his conceptual thought, the author, through his previous extensive experience in multiphoton/tunnel ionization and fragmentation of molecules by intense lasers, proposed that all molecules in the high intensity inside a filament in air would be fragmented and the fragments would fluoresce. He did not consider the physical and chemical processes leading to fluorescence except to assume that after fragmentation, there would be fluorescence. It turns out that the fluorescence is at least partially related to the excitation of the superexcited states of the molecules (Kong et al., 2006; Kong and Chin, 2008). Superexcited states are defined as those molecular states whose total internal energy is higher than the first ionization energy of the molecule. Superexcited states of molecules have been extensively studied by scientists using synchrotron radiation (VUV photons) through one photon absorption (see for example, the review by Hatano, 1999). However, using intense femtosecond laser radiation nonlinearly to excite the superexcited states with temporal resolution is still unexplored. This is in contrast to the excitation by synchrotron radiation which yields only integrated information of the superexcited states. Our recent time-resolved measurement of the superexcited states of CH_4 is just a starting point and would probably lead to a new direction of physical and chemical research in ultrafast intense laser science. More recently, it was observed that the filament zone is birefringent (Béjot et al., 2008; Chen et al., 2008a) because the linearly polarized high intensity pump pulse induced an "instantaneous" anisotropy in the medium. Any probe pulse superposing the pump pulse would "feel" this effect as if it is passing through an anisotropic medium. The pump pulse also excited the rotational wave packet of air molecules so that by delaying the probe pulse, the molecular rotational revival will be observed (Varma et al., 2008; Marceau et al., 2009). These subjects will be covered in the following sections.

S.L. Chin, *Femtosecond Laser Filamentation*, Springer Series on Atomic, Optical, and Plasma Physics 55, DOI 10.1007/978-1-4419-0688-5_8,

8.1 Multiple Filamentation

Although the fundamental physics of single filamentation is mostly understood, multiple filamentation and its control are still challenging subjects. The following expresses some original thoughts.

8.1.1 Why Does a Large Diameter Beam Diverge Slowly Over Long Distances When There Is Multiple Filamentation?

We have explained in Chapter 4 that the slow divergence over long distances is due to the "sucking" of the reservoir energy toward the hot filamenting zones. But there was no explanation as to why this happens. From Chapter 2, the full evolution of a single filament teaches us something. At the end of the filament, the filament core merges with the full beam while the intensity is still rather high, of the order of 10^{10}–10^{11} W/cm^2. Self-focusing (pulling the whole beam toward the center) still takes place but the peak power is lower than the critical power. It is this self-focusing without collapse that slows down the diffraction process so that the beam diverges slowly. If there are multiple filaments, each hot zone will undergo the same process of self-focusing without collapse, pulling the energy around it in the reservoir toward its center of gravity. If competition sets in, each of these hot zones will still propagate with a divergence smaller than that of the linear diffraction although none or very few of them would collapse into a mature filament. Even so, the pulse would propagate very far with a divergence much lower than given by linear diffraction. These hot zones are still rather intense of the order of 10^{10}–10^{11} W/cm^2. They could be used for some specific purposes both constructively and destructively. For example, such hot spots could be sent long distances to hit a transparent target of condensed matter (glass window, etc.). The hot zone would immediately undergo filamentation inside the transparent condensed matter because the critical power of this material is usually very low, of the order of a few MW. Damage might be the consequence. White light generation would be very strong if the whole pulse contains a large amount of energy.

8.1.2 Filament Collaboration

We have discussed in Chapters 4 and 7 the effect of squeezing the filaments to a small zone where constructive interference among the fields of the filaments takes place. This would give rise to a new high intensity zone that keeps on self-focusing. We can now deduce that if there are two filaments that overlap each other, for constructive interference, the resulting intensity would be four times higher than the clamped intensity. If there were three of them overlapping, the resulting intensity would be 9 times the clamped intensity. Imagine now that we could superpose 1,000 filaments together. The resultant constructive interference would give us an increase of intensity by $1{,}000^2$ times, or 10^6 times the clamped intensity; i.e., the resultant intensity would become 5×10^{19} W/cm^2 in air. This is getting into the domain of relativistic intensity. In principle, such experiments could be carried out by tightly focusing petawatt level laser pulses, in air, for example. Unfortunately, this is not the case and would never happen because of intensity clamping as mentioned in chapter 4.6. i.e. in atmospheric density gases (air or argon) the peak intensity attained in the filament shows a remarkable stability, which is independent of the initial pulse energy, geometry of the experiment, single or multiple filament regimes. We prove both experimentally and numerically that neither external focusing of the whole pulse, nor fusion of multiple filaments can lead to the peak intensity increase by more than about 30% of the global maximum in a single filament created by a collimated beam (Kosareva et al, 2009).

8.1.3 Optimum Wavelength to Produce the Broadest and Strongest White Light

When the Ti-sapphire laser pulses self-transforms into a white light laser pulse in air, the broad supercontinuum spans from about 230 nm to 4 μm. On the blue side, the light between 230 and 290 nm comes from the contribution of the third harmonics and its cross phase modulation with the pump pulse. If we ask whether we could extend the wavelength on the UV side toward even shorter wavelengths, we need to understand how efficiently such blue shifts take place. Our current understanding is that the tail end of the blue shift comes from the third harmonic and its cross phase modulation with the pump pulse. It means that we should use a shorter wavelength laser pulse; e.g., 500 nm. But with the short wavelength, the clamped intensity is lower because it takes a lower intensity to ionize the air molecules, resulting in a lower clamped intensity. Since THG is a function of intensity, this lower intensity would mean less efficient generation of TH in the UV side. Hence, there is an optimum wavelength that would give a compromise between broadening the spectrum and the intensity of this broad spectrum. This optimum wavelength is not known yet.

8.1.4 Filament Control Using a Deformable Mirror

So far, using the telescope to focus multiple filaments to long distances has a drawback of astigmatism and other optical aberrations. Recently we have used a deformable mirror to correct the astigmatism and were able to enhance the fluorescence signal from air significantly (Daigle et al., 2008). Figure 8.1 compares the results of corrected and uncorrected focal patterns obtained at a distance of 85 m in air. A better (smaller) focal area is obtained when the wave front is corrected. Figure 8.2 shows the nitrogen fluorescence signal at 337 nm as a function of the pump energy measured at 85 m in air by using a photomultiplier and appropriate

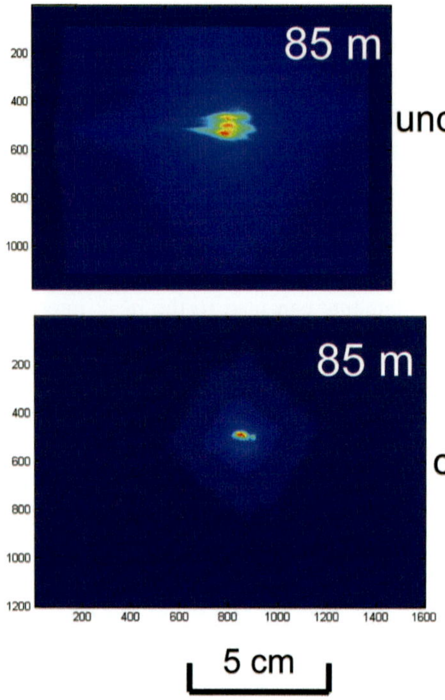

Fig. 8.1 A 800 nm/50 fs laser pulse is focused in air by a telescopic system at 85 m away. The pictures show the reduction in size of the focal pattern when the astigmatism of the wave front of the pulse is corrected

filters to collect the signal from the side of the focal zone. A significant increase of the signal is obtained when the wave front is corrected. However, this result corrected only the astigmatism of the wave front. Other defects such as wave front non-uniformity, turbulence, etc. are not corrected. Since in this type of work, it is the fluorescence signal that needs to be optimized, further corrections should use the fluorescence signal as the correction parameter.

8.2 Time-Resolved Excitation of Superexcited States of Molecules

Synchrotron radiations (SR) in the vacuum ultraviolet with photon energy in the range of 10–50 eV can induce ionization and fragmentation of molecules as well as exciting the superexcited states (Hatano, 1999). The superexcited states will relax through autoionization (coupling with the electronic continuum), neutral dissociation (coupling through the nuclear continuum), photon emission and ion-pair formation. The first two processes have higher transition probabilities (Hatano,

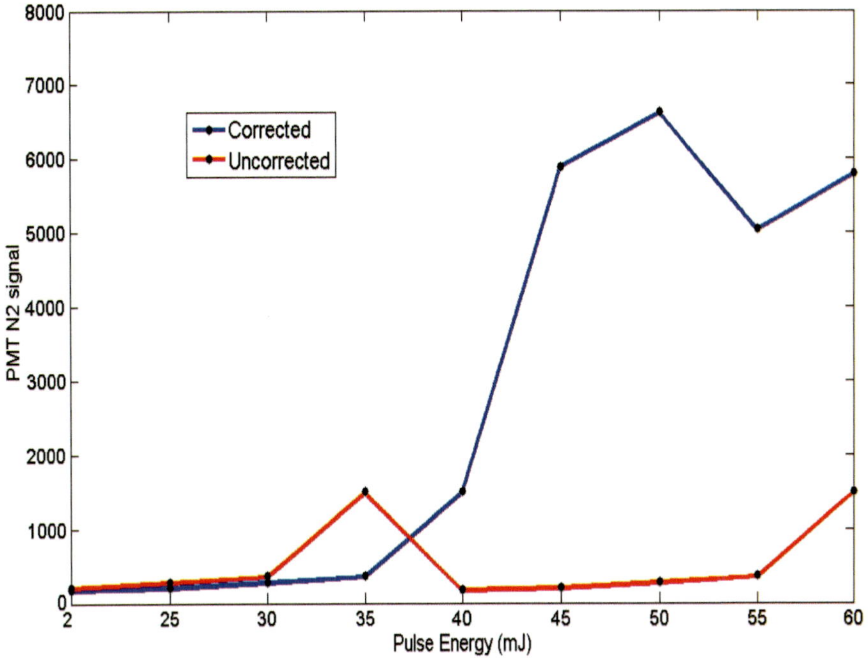

Fig. 8.2 Comparison of nitrogen fluorescence signal at 337 nm as a function of energy of the laser pulse at a distance of 85 m from the telescope. The signal with wave front correction for astigmatism is clearly improved at high pump energy

2001). However, a SR experiment does not provide time-resolved information for the dissociation process of superexcited states. On the other hand, intense femtosecond laser pulses, when interacting with molecules, would lead to multiphoton/tunnel ionization, fragmentation, Coulomb explosion, etc. Excitation of the superexcited states was never considered until recently (Kong et al., 2006; Kong and Chin, 2008). In fact, they realized that the fluorescence spectra from CH_4 excited by synchrotron radiation and by ultrafast intense Ti-sapphire laser pulses are almost identical; i.e., they both yield the fluorescence bands of CH radicals corresponding to the transitions of $A^2\Delta$, $B^2\Sigma^-$, $C^2\Sigma^+ \rightarrow$ X and the Balmer α-line of H ($n = 3$). This similarity can be immediately understood from the energy diagram of Fig. 8.3. Both single VUV photon (SR radiation) absorption or multiphoton/tunnel absorption of intense femtosecond laser pulses would have a definite probability of excitation of the superexcited states of a molecule. Once in a superexcited state, the molecule would either autoionize or would dissociate into neutral fragments (neutral explosion) some of which would be excited; the latter would decay through fluorescence emission. In the case of CH_4, the fluorescence comes from the excited fragments of CH and H.

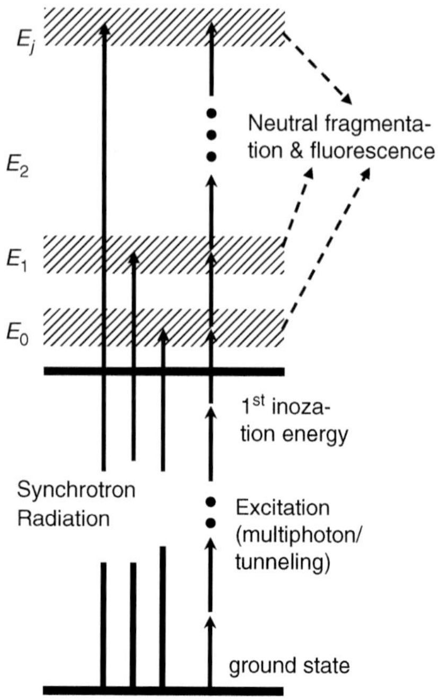

Fig. 8.3 Superexcited states are defined as those molecular states whose total internal energy is higher than the first ionization energy of the molecule. This figure is a schematic energy diagram showing the equivalence of excitation of the superexcited states of a molecule with energy $E_0, E_1, E_2 \ldots E_j$ by synchrotron radiation (one photon absorption) and by intense femtosecond laser pulses (multiphoton/tunnel excitation)

In order to prove this point of view, the author and co-workers (Kong and Chin (2008) and Azarm et al, (2008)) did a pump-probe experiment. The superexcited states, being inside the ionization continuum of the molecule, would couple to the electronic continuum so that its lifetime is very short, most probably in the femtosecond range. They thus used an intense 45 femtosecond Ti-sapphire laser pulse (800 nm) to pump the CH_4 molecules and a 35 femtosecond IR laser pulse (1,330 nm) to probe the superexcited states (Fig. 8.4). It is expected that when the probe pulse arrives within the lifetime of the superexcited states, it would de-excite the superexcited states so that the fluorescence would decrease. Figure 8.5 shows the experimental result. Indeed, there is a sharp decrease of the fluorescence signal within a time of the order of 150 fs. This means that the superexcited state is observed and its lifetime is measured to be about 150 fs.

However, the depth of the dip is only about 5% of the total fluorescence signal. This means that there might be some other mechanism that leads to the fluorescence of CH radicals. We have identified tentatively two other channels; namely, electron-ion recombination and fragmentation of the parent ion.

As already pointed out, this is just the beginning of a new series of studies and much more work is expected to be done in the future by many groups including the author's group.

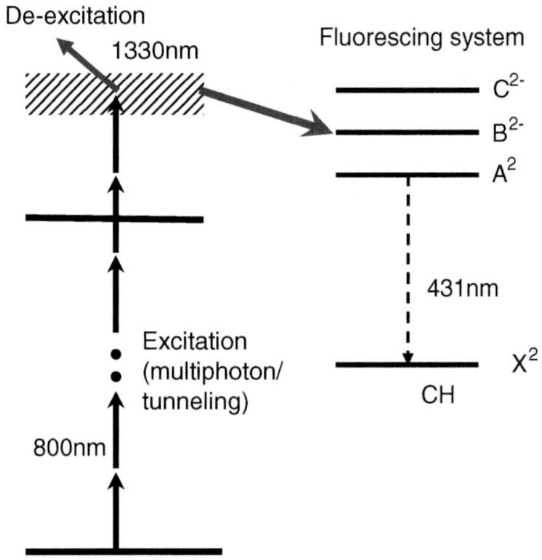

Fig. 8.4 The case of CH_4. The molecules are excited by a 45 fs Ti-sapphire laser pulse to one of its superexcited states which decay to the CH fluorescing system. We monitor the fluorescence at 431 nm. The 35 fs probe pulse de-excites the superexcited state within the latter's lifetime. This would reduce the transfer of population from the superexcited state to the fluorescing CH system; i.e., the fluorescence would decrease

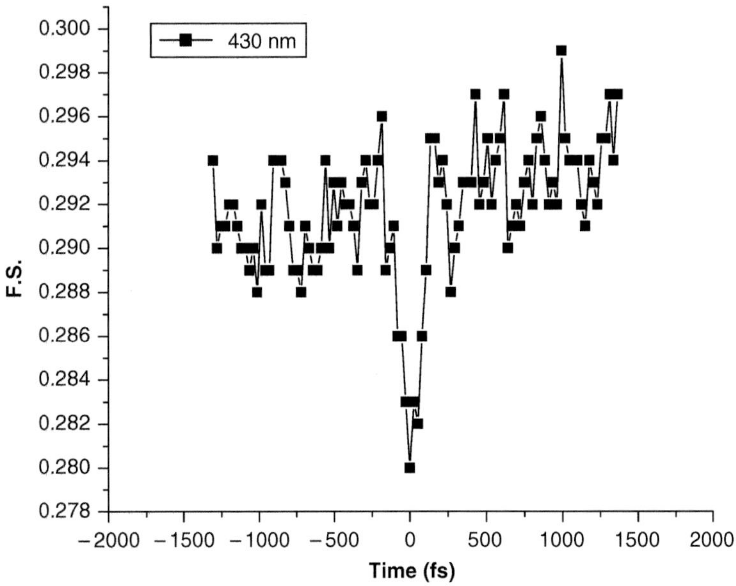

Fig.. 8.5 The fluorescence signal at 431 nm as a function of the delay time between the pump (800 nm) and probe (1330 nm) pulses. Negative delay means that the probe pulse is behind the pump pulse. The width of the dip (FWHM) is of the order of 150 fs

8.3 Ultrafast Birefringence

It was during the past year and a half that the phenomenon of high intensity induced ultrafast birefringence inside the filamenting zone was observed in air and other gases. When a linearly polarized femtosecond laser pulse self-focuses in air during free propagation, the clamped intensity is high: $\sim 5 \times 10^{13}$ W/cm^2. The plasma density is $\sim 10^{16}$ cm^{-3}. This is more than 1,000 times lower than the neutral density, which is about 3×10^{19} cm^{-3}. Thus, most of the medium is still neutral and will experience the effect of the high intensity inside the filament. If the pulse is linearly polarized (which is the case in most experiments), the laser induces a strong material polarization (dipole moment per unit volume) $P_{||}$ parallel to the direction of the laser field's polarization. The isotropic character of air in this zone is thus broken; it becomes a uniaxial medium. Thus, if a femtosecond linearly polarized nonresonant probe pulse of any frequency passes at the same time through this polarized zone with its polarization (electric field) vector making an angle with the polarization (electric field) vector of the filamenting (pump) pulse, it will experience this uniaxial property. That is to say, the index of refraction seen by the probe pulse will be different in the parallel and perpendicular direction with respect to the pump's polarization direction ($n_{||} \neq n_{\perp}$) (Chen et al., 2008a; Béjot et al., 2008). Consequently, the probe pulse becomes elliptically polarized in general during the very short overlapping time with the pump pulse.

Moreover, the strong laser field inside the filament will "instantaneously" induce a rotational wave packet through Raman coupling between the broadband (in frequency) pulse and the rotational modes of the air molecules. That is to say, all the molecules are aligned 'instantaneously' (in a short limited time of the order of tens of femtoseocnds) (Stapelfeldt and Seideman, 2003). This rotational wave packet would dephase and revive. The revival of the rotational wave packet can be observed by the probe pulse through the revival of its polarization after a long delay in the ps regime long after the pump pulse is gone (Calegari et al., 2008; Varma et al., 2008; Marceau et al., 2009). Many periods of revival were observed. More work is expected to be done by various groups in the future.

The birefringence inside the filament leads to the generation of near single cycle tera-hertz (THz) pulses which are elliptically polarized (Chen et al., 2008b). Two-color mixing inside the filament also leads to elliptically polarized THz pulses (Xie et al., 2006; Houard et al., 2008; Chen et al., 2009). The physics of the generation of THz pulses inside filaments is currently a hot subject of debate. This is part of the larger subject of filamentation nonlinear optics. The following discusses in more detail some most recent results of ultrafast birefringence, its application to the easy detection of the excitation and revival of molecular rotational wave packet.

8.3.1 Filament-Induced Birefringence

A 50 fs/800 nm Ti-sapphire laser pulse is focused by a 50 cm lens and generates a filament of about 3 cm long in a gas at atmospheric pressure (air, argon or nitrogen). It is followed collinearly by another roughly 50 fs pulse at 400 nm (second harmonic of the 800 nm fundamental pulse) which acts as a probe at a power significantly

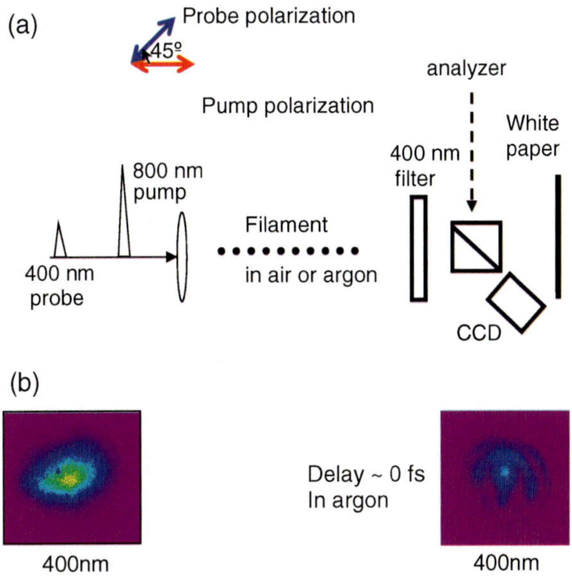

Fig. 8.6 (**a**) Schematic set up of a pump-probe experiment in air or in argon. The probe's linear polarization is set at 45° with respect to the pump's linear polarization. Both pulses pass through a 50 cm lens and the filament length is about 3 cm. The delay between the pump and the probe is adjustable. (**b**) Input and output pattern of the probe pulse when the delay is about 100 fs in air (Chen et al., 2008; Marceau et al., 2009)

lower than the critical power for self-focusing (Fig. 8.6a). The probe can be decomposed into two polarization components, one parallel and one perpendicular to the polarization of the pump. The probe's initial transverse fluence pattern is shown in Fig. 8.6b; left hand picture, when the propagation medium is argon. When there is no delay, the probe's transverse pattern is transformed into a new pattern (Fig. 8.6b, right hand side). This is the pattern observed by Béjot et al. (2008). The probe pulse becomes elliptically polarized.

The mechanism of the observed filament-induced birefringence effect can be explained by the difference in pump-induced nonlinear contributions to the refractive indices along two orthogonal directions as mentioned above. We consider that the two pulses are superimposed onto each other (delay = 0 fs). The probe pulse will feel the birefringence generated by the high intensity of the filamenting pump pulse through cross phase modulation. The latter is an instantaneous electronic response of the atomic gas.

The material polarization experienced by the probe due to cross phase modulation with the pump can be expressed as (Boyd, 2003; Chen et al., 2008a; Béjot et al., 2008)

$$P_{\mathrm{XPM},x}^{\mathrm{probe}} = \frac{3\varepsilon_0}{2}\chi_{xxxx}^{(3)}\left|E_x^{\mathrm{filament}}\right|^2 E_x^{\mathrm{probe}} \tag{8.1}$$

$$P_{\mathrm{XPM},y}^{\mathrm{probe}} = \frac{3\varepsilon_0}{2}\chi_{yyxx}^{(3)}\left|E_x^{\mathrm{filament}}\right|^2 E_y^{\mathrm{probe}} \tag{8.2}$$

where x denotes the direction of the pump polarization and E_x^{filament} is the electric field inside the filament. E_x^{probe} and E_y^{probe} are the electric fields of the probe pulse polarized in the x- *and* y-directions respectively. $\chi_{xxxx}^{(3)}$, $\chi_{yyxx}^{(3)}$ are the corresponding third order susceptibility tensors.

$$\begin{aligned}&\chi_{ijkl}^{(3)} \equiv \chi_{ijkl}^{(3)}\left(-\omega_i^{\text{probe}};\, \omega_j^{\text{probe}},\, \omega_k^{\text{filament}},\, -\,\omega_l^{\text{filament}}\right)\\ &i,j,k,l(=x,\, y)\text{: directions of the electric fields}\\ &\qquad\qquad\qquad\text{of the corresponding frequencies}\end{aligned} \tag{8.3}$$

The index of refraction, n, experienced by the probe pulse, according to normal electromagnetic theory, is given in general by the following consideration in scalar form in which P and E correspond to the total material polarization and the electric field of the probe, respectively:

$$\begin{aligned}&P = \varepsilon_0 \chi E\\ &D \equiv \varepsilon E = P + \varepsilon_0 E = \varepsilon_0(1+\chi)E\\ &\varepsilon = \varepsilon_0(1+\chi)\\ &n^2 \equiv \varepsilon/\varepsilon_0\\ &n = (1+\chi)^{1/2} = \left(1 + \frac{P}{\varepsilon_0 E}\right)^{1/2}\end{aligned} \tag{8.4}$$

D is the electric flux density in the material and χ is the susceptibility. Explicitly, the index depends on the direction of the polarization and is given by

$$n^{\text{probe}}(i) = \left[1 + \frac{P_{\text{total}}^{\text{probe}}}{\varepsilon_0 E_i^{\text{probe}}}\right]^{1/2} = \left[1 + \frac{P_0^{\text{probe}}}{\varepsilon_0 E_i^{\text{probe}}} + \frac{P_{\text{XPM},i}^{\text{probe}}}{\varepsilon_0 E_i^{\text{probe}}}\right]^{1/2} ;i = x,y \tag{8.5}$$

n^{probe}(i): total index of refraction of the probe E_i^{probe}
P_0^{probe}: linear polarization of the probe (isotropic)

We define n_0 = linear index of refraction (isotropic); hence,

$$n_0^2 \equiv 1 + \frac{P_0^{\text{probe}}}{\varepsilon_0 E_i^{\text{probe}}} \quad (\text{for all } i,\ \because \text{ isotropic}) \tag{8.6}$$

and

$$\therefore\ n^{\text{probe}}(i) = \left[n_0^2 + \frac{P_{\text{XPM},i}^{\text{probe}}}{\varepsilon_0 E_i^{\text{probe}}}\right]^{1/2} ;\quad i = x,y \tag{8.7}$$

Simplifying,

$$
\begin{aligned}
n^{\text{probe}}(i) &= n_0\left[1+\frac{P_{\text{XPM},i}^{\text{probe}}}{n_0^2\varepsilon_0 E_i^{\text{probe}}}\right]^{1/2} \cong n_0\left[1+\frac{P_{\text{XPM},i}^{\text{probe}}}{2n_0^2\varepsilon_0 E_i^{\text{probe}}}\right] \\
&= n_0 + \frac{P_{\text{XPM},i}^{\text{probe}}}{2n_0\varepsilon_0 E_i^{\text{probe}}} \quad (i = x,y)
\end{aligned} \tag{8.8}
$$

We define

$$
\begin{aligned}
\Delta \mathrm{n}_{\|} &\equiv n^{\text{probe}}(x) - n_0 \\
\Delta n_{\perp} &\equiv n^{\text{probe}}(y) - n_0
\end{aligned} \tag{8.9}
$$

$$
\begin{aligned}
\Delta \mathrm{n}_{\|} &= \frac{P_{\text{XPM},x}^{\text{probe}}}{2n_0\varepsilon_0 E_x^{\text{probe}}} \\
&= \frac{3\varepsilon_0}{4n_0}\mathrm{Re}(\chi_{xxxx}^{(3)})\left|E_x^{\text{filament}}\right|^2 \ (\text{from Eq. 8.1})
\end{aligned} \tag{8.10}
$$

and

$$
\begin{aligned}
\Delta n_{\perp} &= \frac{P_{\text{XPM},y}^{\text{probe}}}{2n_0\varepsilon_0 E_y^{\text{probe}}} \\
&= \frac{3\varepsilon_0}{4n_0}\mathrm{Re}(\chi_{yyxx}^{(3)})\left|E_x^{\text{filament}}\right|^2 \ (\text{from Eq. 8.2})
\end{aligned} \tag{8.11}
$$

Hence,

$$
\therefore \ \frac{\Delta \mathrm{n}_{\|}}{\Delta n_{\perp}} = \frac{\mathrm{Re}\,(\chi_{xxxx}^{(3)})}{\mathrm{Re}\,(\chi_{yyxx}^{(3)})} \tag{8.12}
$$

Since, for an isotropic medium (Boyd, 2003)

$$
\mathrm{Re}(\chi_{xxxx}^{(3)}) = 3\ \mathrm{Re}\,(\chi_{yyxx}^{(3)}) \tag{8.13}
$$

we have

$$
\Delta n_{\|} = 3\Delta n_{\perp} \tag{8.14}
$$

That is to say, the nonlinear index change (increase) of the probe with parallel polarization, $\Delta n_{\|}$, is larger than that with perpendicular polarization $\Delta n_{\perp}$. Since both nonlinear indices of refraction are positive, the two polarization components of the probe are guided in the filament zone but each will have a different velocity of propagation. Looking at the pattern at the right hand side of Fig. 8.6b, there is a central spot surrounded by some ring-like structure. This would mean that the central spot is the superposition of the parallel and perpendicular components of the probe pulse due to guiding through their respective positive nonlinear indices. But since the two indices are different and there is a weak plasma diffraction inside the filament zone, the parallel-to-pump component might always be guided at the center with some radial spreading while the perpendicular-to-pump component is diffracted more giving rise to the ring. It will however contribute also to the central spot resulting in a complicated polarization distribution. The probe pattern becomes elliptically polarized but the ellipticity is evidently not uniform across the pattern. More detailed experiments need to be done for a better understanding of the phenomenon.

8.3.2 Excitation of Molecular Rotational Wave Packets in Air and Polarization Separation

The above observation of the filament-induced birefringence is just one aspect of filament-induced anisotropy through the instantaneous electronic response of argon atoms. This electronic response is universal for all media, including molecular gases when the filamenting pump pulse and the probe pulse are superimposed (delay $= 0$ fs). Meanwhile, there is another phenomenon that shows up in the filamentation process when the medium is a molecular gas such as air. Air molecules in the high intensity of the filament zone could be excited rotationally due to the Raman transition. Because the fs laser pulse has a broadband width, many rotational modes of the molecules are excited simultaneously resulting in the excitation of a rotational wave packet (Seideman, 1995; Seideman and Hamilton, 2006). The molecular wave packet is initially aligned in the direction of the pump's linear polarization. Such molecular wave packets, once formed inside the filament zone, will start to dephase and later revive back into the original wave packet in the picosecond time scales, long after the pump fs pulse is gone. When the probe pulse is delayed, it will sample the rotational wave packet at its various states of dephasing and revival (Calegari et al., 2008; Varma et al., 2008). The observation of such rotational wave packet in high pressure molecular gas is interesting because it is easily observable as compared to previous work done in high vacuum systems looking at individual molecules. Also interesting is the fact that the revival does not decay quickly at atmospheric pressure through collision, of the order of 100 ps (Marceau et al., 2009).

Let us consider the excitation of the rotational wave packet in air. Initially, the air molecules (nitrogen and oxygen) are excited rotationally through Raman scattering by the broadband 800 nm Ti-sapphire laser pulse. The excitation is a quantum mechanical electronic process; hence, it is an instantaneous process. But this does not mean that all the molecules are immediately aligned in the pump's polarization direction. It takes some response time to have the wave packet build up into one with all the molecules aligned before it de-phases out. This build up time is of the order of 100 fs according to our observation. Let us see how this is observed.

We refer back to Fig. 8.6a for the pump-probe experiment comparing now argon gas, air and nitrogen. Figure 8.7 shows the results. In the case of argon gas (see top row in Fig. 8.7a), the probe pattern first appears at delay zero. This is the same pattern shown in Fig. 8.6b (right hand pattern). As the delay increases, this pattern disappears quickly once the pump and probe no longer overlap each other. In the

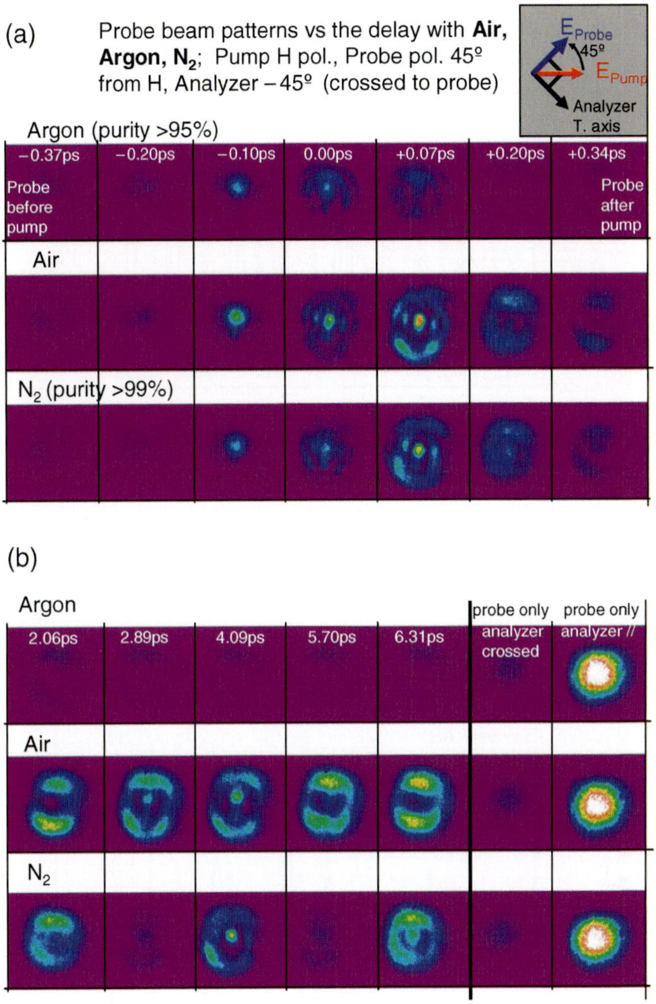

Fig. 8.7 Revival of the probe patterns in air and nitrogen but not in argon after the pump pulse is long gone (many ps later)

case of air and nitrogen, the initial patterns at zero delay are similar to that of argon. However, these patterns do not disappear as the delay increases. Rather, they disappear at delays long after the pump is gone and reappear again at different intervals (Fig. 8.7a and b). If we now integrate the pattern's energy content, we obtain the plot shown in Fig. 8.8 showing the transmitted probe energy vs. delay. We now make a close-up at the first few revivals as shown in Fig. 8.9. The argon curve peaks very early and decays soon after. We define this argon peak as the zero delay. The height of the peak corresponds to the peak transmission of the probe pulse; i.e., the interaction is strongest. Strongest interaction implies best alignment of the molecules

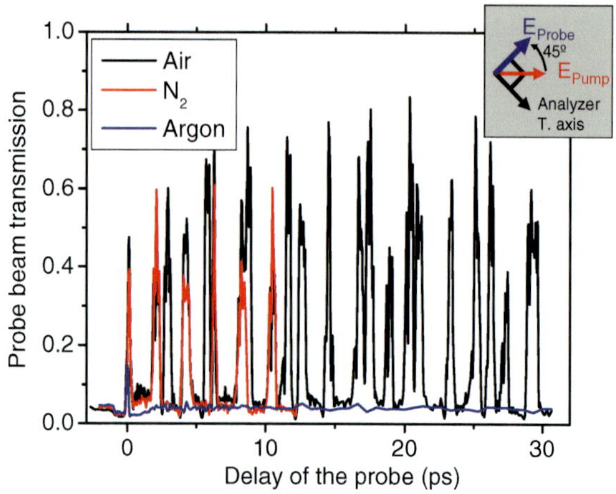

Fig. 8.8 The probe patterns are integrated giving the transmitted energy of the probe through the filament zone. As the delay is changed, the probe's transmitted energy changes according to the revival of nitrogen's and oxygen's rotational wave packet. Argon does not have revival as expected. The full revival periods for nitrogen and oxygen are 8.2 and 11.6 ps while we measured 8.4 and 11.6 ps, respectively. In the plot above, the signal through nitrogen was scanned till about 12 ps. For oxygen, it was scanned up to 30 ps. Collisional de-coherence of the wave packet in one atmospheric gas takes about 100 ps

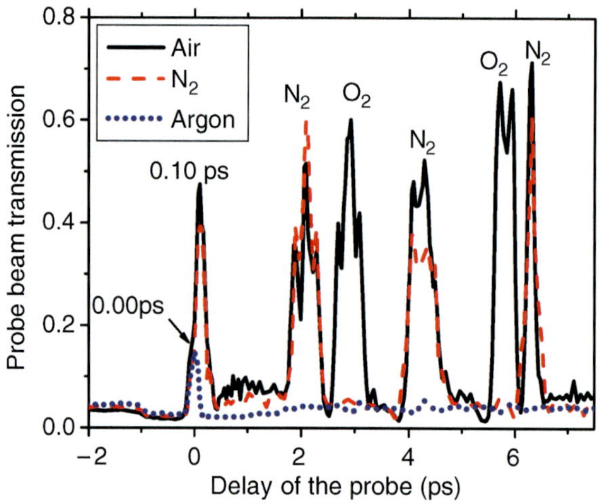

Fig. 8.9 Close-up of the first few revival curves of Fig. 8.8. Height of the peaks correspond to the transmission of the probe pulse. At the peak, the transmission is highest; i.e., the interaction is strongest. Strongest interaction implies best alignment of the molecules in the wave packet. Thus the peak after the argon peak (defined as zero delay) is where all the molecules are aligned. This peak is at a delay of about 100 fs. We do not have enough resolution to distinguish the initial alignment times of oxygen and nitrogen. But later their revival times are distinguishable as indicated by the identifications next to each peak. The relative intensity of the different peaks should not be quantitatively compared since, as we found out later, the transmission bandwidth of the band pass filter was narrower than the spectral broadening (Calegari et al., 2008) of the probe pulse at different delays. Broader band pass filters were used to confirm that our conclusions were still valid

in the wave packet. Thus the peak of the molecular gases (air and nitrogen) after the argon peak is where all the molecules are aligned. This peak is at a delay of about 100 fs. Experiments up to now have not had enough resolution to distinguish the initial alignment times of oxygen and nitrogen. But later their revival times are distinguishable as indicated by the identifications next to each peak in Fig. 8.9. The relative intensity of the different peaks should not be quantitatively compared since the transmission bandwidth of the band pass filter of the probe beam was narrower than the spectral broadening (Calegari et al., 2008) of the probe pulse at different delays. Broader band pass filters were used to confirm that the conclusions were still valid.

When we measure the probe's pattern at a delay around the first peak of air in Fig. 8.9, we obtain the pattern shown in Fig. 8.10. Using an analyzer to check the detail of the polarization shows that the outer ring's linear polarization is perpendicular to that of the inner pattern (Fig. 8.10). The inner pattern has a linear polarization parallel to that of the pump. Thus, the initial beam pattern of the probe beam is separated into two distinct parts of orthogonal polarizations. This phenomenon is called "polarization separator" as compared to the mixed state when the pump and probe pulses are superimposed (Fig. 8.7) at zero delay time. Detailed experimental proof of the above phenomenon can be found in Chen et al. (2008a). This polarization separation effect is due to the different indices of refraction seen by the probe in the parallel and perpendicular to the pump polarization direction. We note that the refractive index modulation (Δn) of the probe pulse is proportional to ($<\cos^2\theta>-1/3$), where θ is the angle between the molecular axis of the rotational wave packet and the polarization direction of the probe and "< >" means the average value (Calegari et al., 2008; Varma et al., 2008). Hence, $\Delta n_{||}(\tau)$ is positive for the parallel-to-pump component ($\theta = 0$), while $\Delta n_{\perp}(\tau)$ is negative for the perpendicular-to-pump component ($\theta = 90°$). Thus, after the delay of $\sim$100 fs when all the molecules are aligned, the parallel index change which is positive will guide the probe resulting in the central part of the pattern shown in Fig. 8.10. This central part consists of a central spot

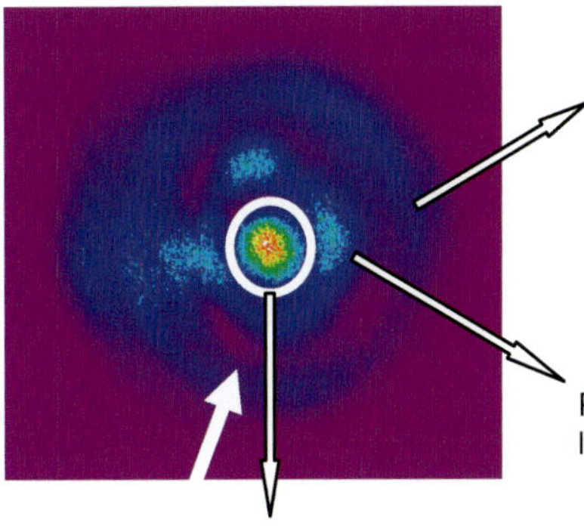

Fig. 8.10 Detailed measurement of the polarization of the pattern using an analyzer shows that the *outer ring* and the *inner pattern* are orthogonally polarized, the *outer ring* being perpendicular to the pump polarization while the *inner part*, parallel. The delay between the pump and probe is of the order of 100 fs (Chen et al., 2008)

and a "flower" pattern. We believe that the flower pattern might be due to either the diffraction by the plasma inside the filament or the elliptical shape of the initial pattern. The latter would result in a symmetrically located hot spot during filamentation (See chapter 4, section 4.3.). More work needs to be done to clarify this. The perpendicular component of the probe beam sees the negative indices of both the aligned molecules and the plasma. It is thus strongly diffracted resulting in the outer ring.

We can now also understand why there is a revival in the signal shown in Figs. 8.7, 8.8, and 8.9. The dephasing of the rotational wave packet is followed by revival. When the revival happens, the molecules are aligned again and the probe beam sees the change of index. It will guide the parallel polarization component of the probe at the center and diffract the perpendicular component periodically. The full revival periods for nitrogen and oxygen are 8.2 and 11.6 ps, while we measured 8.4 and 11.6 ps, respectively.

8.3.3 *Just the Beginning of Filament-Induced Birefringence*

The above recent observation of filament-induced birefringence is just the beginning of what could become another chapter on filamentation. There is a need for theoretical simulation. In particular, all current theories assumed that the filament zones remain isotropic during propagation while in fact, they become anisotropic. What would be the influence of such birefringence effect on the pulse itself? If the pulse becomes longer, say, one ps, the molecular alignment effect would have already set in, in the case of air, for example. What would be the consequence? Other consequences in applications are not yet known. The field is thus wide open for innovation.

References

Aközbek N, Bowden CM, Talebpour A and Chin SL (2000) Phys Rev E 61:4540.
Aközbek N, Scalora M, Bowden CM and Chin SL (2001) Opt Commun 191:353.
Aközbek N, Iwasaki A, Becker A, Scalora M, Chin SL and Bowden CM (2002a) Phys Rev Lett 89:143901.
Aközbek N, Bowden CM and Chin SL (2002b) J Mod Opt 49:475.
Aközbek N, Becker A, Scalora M, Chin SL and Bowden CM (2003) Appl Phys B 77:177.
Alfano RR (ed.) (1989) The Supercontinuum Laser Source, Springer Verlag, Berlin.
Arevalo E and Becker A (2005) Phys Rev E 72:026605.
Akturk S, D'Amico C, Franco M, Couairon A and Mysyrowicz A (2007) Phys Rev A 76:063819.
Augst S, Meyerhofer DD, Strickland D and Chin SL (1991) JOSA-B 8:858–867.
Azarm A, Xu HL, Kamali Y, Bernhardt J, Song D, Xia A, Teranishi Y, Lin SH, Kong F, and Chin SL (2008) Direct observation of super-excited states in methane created by a femtosecond intense laser field, J Phys B:At Mol & Opt Phys, 41:225601.
Becker A, Akozbek N, Vijayalakshmi K, Oral E, Bowden CM and Chin SL (2001) Appl Phys B 73:287–290.
Becker A, Bandrauk AD, Chin SL (2001a) Chem Phys Lett 343:345.
Becker A, Plaja L, Moreno P, Nurhuda M and Faisal FHM (2001b) Phys Rev A 64:023408.
Becker W, Grasbon F, Kopold R, Milosevic DB, Paulus GG and Walther H (2002) Adv At Mol Opt Phys 48:35–98.
Becker A and Faisal FHM (2005) J Phys B 38:R1.
Béjot P, Bonacina L, Extermann J, Moret M, Wolf JP, Ackermann R, Lascoux N, Salamé R, Salmon E, Kasparian J, Bergé L, Champeaux S, Guet C, Blanchot N, Bonville O, Boscheron A, Canal P, Castaldi M, Hartmann O, Lepage C, Marmande L, Mazataud E, Mennerat G, Patissou L, Prevot V, Raffestin D and Ribolzi J (2007) Appl Phys Lett 90:151106.
Béjot P, Petit Y, Bonacina L, Kasparian L, Moret M and Wolf J-P (2008) Opt Exp 16:7564.
Bergé L, Skupin S, Lederer F, Meéjean G, Yu J, Kasparian J, Salmon E, Wolf JP, Rodriguez M, Woeste L, Bourayou R and Sauerbrey R (2004a) Phys Rev Lett 92:225002-1–225002-4.
Bergé L, Skupin S, Lederer F, Méjean G, Yu J, Kasparian J, Salmon E and Wolf JP (2004b) Phys Scrip T107:135.
Bergé L, Skupin S, Nuter R, Kasparian J and Wolf J-P (2007) Rep Prog Phys 70:1633.
Bernhardt J, Liu W, Chin SL and Sauerbrey R (2008) Appl Phys B 91:45.
Bespalov VI and Talanov VI (1966) Sov Phys JETP Lett 3:307.
Boyd RW (2003) Nonlinear Optics, 2nd edition, Academic Press, Amsterdam.
Braun A, Korn G, Liu X, Du D, Squier J and Mourou G (1995) Opt Lett 20:73.
Brodeur A, Chien CY, Ilkov FA, Chin SL, Kosareva OG and Kandidov VP (1997) Opt Lett 22:304.
Brodeur A and Chin SL (1998) Phys Rev Lett 80:4406.
Brodeur A and Chin, SL (1999) J Opt Soc Am B 16:637.
Calegari F, Vozzi C, Gasilov S, Benedetti E, Sansone G, Nisoli M, De Silvestri S and Stagira S (2008) Phys Rev Lett 100:123006.

S.L. Chin, *Femtosecond Laser Filamentation*, Springer Series on Atomic, Optical, and Plasma Physics 55, DOI 10.1007/978-1-4419-0688-5,

Carrasco S, Polyakov S, Kim H, Jankovic L and Stegeman G (2003) Phys Rev E 67:046616.
Châteauneuf M, Payeur S, Dubois J and Kieffer J-C (2008) Appl Phys Lett 92:091104.
Chen Y, Théberge F, Kosareva O, Panov N, Kandidov VP and Chin SL (2007) Opt Lett 32:3477.
Chen Y, Marceau C, Théberge F, Châteauneuf M, Dubois J and Chin SL (2008a) Opt Lett 33:2731.
Chen Y, Marceau C, Liu W, Sun Z-D, Zhang Y, Théberge F, Châteauneuf M, Dubois J and Chin SL (2008b) Appl Phys Lett 93:231116.
Chen Y, Marceau C, Génier S, Théberge F, Châteauneuf M, Dubois J and Chin SL (2009) Opt Comm, 282:4283.
Chiao RY, Garmire E and Townes CH (1964) Phys Rev Lett 13:479.
Chin SL (1970) Can J Phys 48:1314.
Chin SL (1984) Chapter 1, Introduction. In: Multiphoton Ionization of Atoms, Chin SL and Lambropoulos P (eds.), Academic Press, Toronto.
Chin SL (2006) J Korean Phys Soc 49:281.
Chin SL (2007) Method for remote sensing of pollutant molecules in a transparent medium using ultrashort intense lasers, US patent, Feb. 27, 2007, No. US 7,184,143 B2.
Chin SL and Isenor N (1967) Phys Rev 158:93–94.
Chin SL and Lambropoulos P (eds.) (1984) Multiphoton Ionization of Atoms, Academic Press, Toronto.
Chin SL, Yergeau Y and Lavigne P (1985) J Phys B 18:L213.
Chin SL, Francois V, Watson JM and Delisle C (1992) Appl Opt 31:3383.
Chin SL, Brodeur A, Petit S, Kosareva OG and Kandidov VP (1999) J Nonlinear Opt Phys Mater 8:121.
Chin SL, Akozbek N, Proulx A, Petit S, Bowden CM (2001) Opt Commun, 188:181.
Chin SL, Petit S, Liu W, Iwasaki A, Nadeau M-C, Kandidov VP, Kosareva OG and Andrianov KY (2002) Opt Commun 210:329.
Chin SL (2004) From multiphoton to tunnel ionization. In: Progress in Multiphoton Processes and Spectroscopy, Lin SH, Fujimura Y and Villaeys AA (eds.), World Scientific, Singapore, NJ, London, Hong Kong, vol. 16, pp. 273–306.
Chin SL, Hosseini SA, Liu W, Luo Q, Théberg F, Aközbek N, Becker A, Kandidov VP, Kosareva OG and Schroeder H (2005) Can J Phys 83:863–905.
Chin SL, Théberge F, Liu W (2007) Filamentation nonlinear optics. Appl Phys B 86:477.
Chin SL, Liu W, Théberge F, Luo Q, Hosseini SA, Kandidov VP, Kosareva OG, Akozbek N, Becker A, Schroeder H (2008a) Some fundamental concepts of femtosecond laser filamentation. In: Progress in Ultrafast Intense Laser Science III, Springer Series in Chemical Physics, Yamanouchi K, Chin SL, Agostini P and Ferrante G (eds.), Springer, Berlin, vol. 3, Chapter 12, p. 243.
Chin SL, Chen Y, Kosareva O, Kandidov VP and Théberge F, (2008b) What is a filament? Laser Phys 18:962.
Chin SL, Xu HL, Luo Q, Théberge F, Liu W, Daigle J-F, Kamali Y, Simard PT, Bernhardt J, Hosseini SA, Sharifi M, Méjean G, Azarm A, Marceau C, Kosareva O, Kandidov VP, Aközbek N, Becker A, Roy G, Mathieu P, Simard J-R, Châteauneuf M and Dubois J (2009) Appl Phys B 95:1.
Chiron A, Lamoroux B, Lange R, Ripoche J-F, Franco M, Prade B, Bonnaud G, Riazuelo G and Mysyrowicz A (1999) Eur Phys J D 6:383.
Comtois D, Pépin H, Vidal F, Rizk FAM, Chien CY, Johnston TY, Kieffer J-C, La Fontaine B, Martin F, Potvin C, Couture P, Mercure HP, Bondiou-Clergerie A, Lalande P and Gallimberti I (2003) IEEE Trans Plasma Sci 31:377–386 and 387–395.
Conti C (2005) Nonlinear x-waves or x-solitons, lecture note, cf: the following website: http://nlo.phys.uniroma1.it/TalkNLXAquila2005.pdf
Corkum PB (1993) Phys Rev Lett 71:1994.
Couairon A, Franco M, Mysyrowicz A, Biegert J and Keller U (2005) Opt Lett 30:2657.
Couairon A, Gaižauskas E, Faccio D, Dubietis A and Di Trapani P (2006) Phys Rev E 73:016608.
Couairon A and Mysyrowicz A (2007) Phys Rep 441:47–189.

Courvoisier F, Boutou V, Kasparian J, Salmon E, Méjean G, Yu J and Wol J-P (2003) Appl Phys Lett 83:123.
Daigle J-F, Méjean G, Liu W, Théberge F, Xu HL, Kamali Y, Bernhardt J, Azarm A, Sun Q, Mathieu P, Roy G, Simard J-R and Chin SL (2007a) Appl Phys B 87:749.
Daigle J-F, Mathieu P, Roy G, Simard J-R and Chin SL (2007b) Opt Commun 278:147.
Daigle J-F, Kamali Y, Bernhardt J, Liu W, Marceau C, Azarm A and Chin SL (2008) Opt Commun 281:3327.
Delone NB and Krainov VP (1994) Multiphoton Processes in Atoms, Springer-Verlag, Berlin.
Diels J-C and Rudolph W (2006) Ultrashort Laser Pulse Phenomenoa: Fundamentals, Techniques, and Applications on a Femtosecond Time Scale, Academic Press, Burlington, MA.
Di Trapani P, Valiulis G, Piskarskas A, Jedrkiewicz O, Trull J, Conti C and Trillo S (2003a) Phys Rev Lett 91:093904.
Di Trapani P, Valiulis G, Piskarskas A, Jedrkiewicz O, Trull J, Conti C and Trillo S (2003b) Phys Rev Focus, 4 September, 2003.
Dormidonov AE, Valuev VV, Dmitriev VL, Shlenov SA and Kandidov VP (2007) Proc SPIE 6733:673325.
Dubietis A, Tamosauskas G, Fibish G and Ilan B (2004a) Opt Lett 29:1126.
Dubietis A, Gaiz¡auskas E, Tamos¡auskas G and Trapani PD (2004b) Phys Rev Lett 92:253903.
Eisenmann S, Pukhov A and Zigler A (2007) Phys Rev Lett 98:155002.
Fehér M and Martin PA (1995) Spectrochim Acta Part A 51:1579.
Fiji T, Horio T and Suzuki T (2007) Generation of intense deep UV 10 fs pulses by 4WM through filamentation in gases, CLEO 2007, Technical digest, CTuW6.
Gaeta AL (2000) Phys Rev Lett 84:3582.
Goeppert-Mayer M (1931) Ann Phys (Leipzig) 9:273.
Golubtsov IS, Kandidov VP and Kosareva OG (2001) Atmos Oceanic Opt 14:303.
Gravel J-F, Luo Q, Boudreau D, Tang XP and Chin SL (2004) Anal Chem 76:4799.
Hauri CP, Kornelis W, Helbing FW, Heinrich A, Couairon A, Mysyrowicz A, Biegert J and Keller U (2004) Appl Phys B 79:673.
Hatano Y (1999) Phys Rep 313:109.
Hatano Y (2001) J Electron Spectrosc Relat Phenom 119:107.
Hercher M (1964) JOSA 54:563.
Hosseini SA, Luo Q, Ferland B, Liu W, Aközbek N, Roy G and Chin SL (2003) Appl Phys B 77:697.
Hosseini SA, Luo Q, Ferland B, Liu W, Chin SL, Kosareva OG, Panov NA, Aközbek N and Kandidov VP (2004) Phys Rev A 70:033802.
Houard A, Liu Y, Prade B and Mysyrowicz A (2008) Opt Lett 33:1195.
Kamali Y, Sun Q, Daigle J-F, Azarm A, Bernhardt J and Chin SL (2009a) Opt Comm, 282:950.
Kamali Y, Daigle J-F, Théberge F, Châteauneuf M, Azarm A, Chen Y, Marceau C, Lessard SC, Lessard F, Roy G, Dubois J, and Chin SL (2009b) Opt Comm, 282:2062.
Kandidov VP, Kosareva OG, Golubtsov IS, Liu W, Becker A, Akõzbek N, Bowden CM and Chin SL (2003a) Appl Phys B 77:149.
Kandidov VP, Kosareva OG and Koltun AA (2003b) Quant Electron 33:69.
Kandidov VP, Golubtsov IS and Kosareva OG (2004) Quant Electron 34:348.
Kandidov VP, Akozbek N, Scalora M, Kosareva OG, Nyakk AV, Luo Q, Hosseini SA and Chin SL (2005) Appl Phys B 80:267.
Kasparian J, Sauerbrey R and Chin SL (2000a) Appl Phys B 71:877.
Kasparian J, Sauerbrey R, Mondelain D, Niedermeier S, Yu J, Wolf J-P, Andre Y-B, Franco M, Prade B, Tzortzakis S, Mysyrowicz A, Rodriguez M, Wille H and Woeste L (2000b) Opt Lett 25:1397.
Kasparian J, Rodriguez M, Méjean G, Yu J, Salmon E, Wille H, Bourayou R, Frey S, André Y-B, Mysyrowicz A, Sauerbrey R, Wolf J-P and Woeste L (2003) 301:61.
Kasparian J and Wolf J-P (2008) Opt Exp 16:466.

Keldysh LV (1965) Sov Phys JETP 20:1307. (Russian original: Keldysh, LV (1964) Zh Eksp Teor Fiz 47:1945.)
Kolesik M and Moloney JV (2004) Self-healing femtosecond light filaments. Opt Lett 29:590.
Kolesik M, Wright EM and Moloney JV (2004) Phys Rev Lett 92:253901.
Kong F, Luo Q, Xu HL, Sharifi M, Song D and Chin SL (2006) J Chem Phys 125:133320.
Kong F and Chin SL (2008) Non-coulomb explosions of molecules in intense laser fields. In: Progress in Ultrafast Intense Laser Science II, Springer Series in Chemical Physics, Yamanouchi K, Chin SL, Agostini P and Ferrante (eds.), Springer, Berlin, vol. 3, Chapter 6, p. 113.
Koprinkov IG, Suda A, Wang P and Midorikawa K (2000) Phys Rev Lett 84:3847.
Kosareva OG, Kandidov VP, Brodeur A and Chin SL (1997a) J Nonlinear Opt Phys Mater 6:485.
Kosareva OG, Kandiddov VP, Brodeur A, Chien CY and Chin SL (1997b) Opt Lett 22:1332.
Kosareva OG, Panov NA, Akozbek N, Kandidov VP, Luo Q, Hosseini SA, Liu W, Gravel J-F, Roy G and Chin SL (2006) Appl Phys B 82:111.
Kosareva OG, Murtazin IN, Panov NA, Savel'ev AB, Kandidov VP and Chin SL (2007) Laser Phys Lett 4:126.
Kosareva OG, Panov NA, Uryupina DS, Kurilova MV, Mazhorova AV, Savel'ev AB, Volkov RV, Kandidov VP and Chin SL (2008) Appl Phys B 91:35.
Kosareva OG, Liu W, Panov NA, Bernhardt J, Ji Z, Sharifi M, Li R, Xu Z, Liu J, Wang Z, Ju J, Lu X, Jiang Y, Leng Y, Liang X, Kandidov VP, and Chin SL (2009) Laser Phys 19:1776.
Kuchiev M Yu (1987) Sov Phys JETP Lett 45:404.
Kuchiev M Yu (1995) J Phys B At Mol Opt Phys 28:5093.
Kuchiev M Yu (1996) Phys Lett A 212:77.
La Fontaine B, Vidal F, Jiang Z, Chien CY, Corntois D, Desparois A, Johnston TW, Kieffe J-C, Pepin H & Mercure HP (1999) Physics of Plasmas 6:1615.
Liu W and Chin SL (2005) Opt Exp 13:5750.
Liu W, Luo Q and Chin SL (2003) Chin Opt Lett 1:56.
Liu W, Petit S, Becker A, Aközbek N, Bowden CM and Chin SL (2002) Opt Commun 202:189.
Liu W, Hosseini SA, Luo Q, Ferland B, Chin SL, Kosareva OG, Panov NA and Kandidov VP (2004) New J Phys 6:1–22.
Liu W, Gravel J-F, Théberge F, Becker A and Chin SL (2005a) Appl Phys B 80:857.
Liu W, Théberge F, Arévalo E, Gravel J-F, Becker A and Chin SL (2005b) Opt Lett 30:2602.
Liu W, Luo Q, Théberge F, Xu H, Hosseini SA, Sharifi SM and Chin SL (2006a) Appl Phys B 82:373.
Liu W, Théberge F, Daigle J-F, Simard PT, Sarifi SM, Kamali Y, Xu HL and Chin SL (2006b) Appl Phys B 85:55.
Liu W and Chin SL (2007) Phys Rev A 76:013826.
Liu W, Xu HL, Méjean G, Kamali Y, Daigle J-F, Azarm A, Simard PT, Mathieu P, Roy G and Chin SL (2007) Spectrochimica Acta Part B 62:76.
Luo Q, Hosseini SA, Ferland B and Chin SL (2004) Opt Commun 233:411.
Luo Q, Hosseini SA, Liu W, Gravel J-F, Kosareva OG, Panov NA, Aközbek N, Kandidov VP, Roy G and Chin SL (2005) Appl Phys B 80:35.
Luo Q, Xu HL, Hosseini SA, Daigle J-F, Theberge F, Sharifi M, Chin SL (2006) Appl Phys B 82:105.
Luther GG, Newell AC, Moloney JV and Wright EM (1994) Opt Lett 19:789.
Marburger JH (1975) Prog Quant Electron 4:35.
Marceau C, Chen Y, Théberge F, Châteauneuf M, Dubois J and Chin SL (2009) Opt Lett 34:1417.
Measures RM (1992) Laser Remote Sensing, Fundamentals and Applications, Krieger Publishing Co., Malabar, FL (reprinted edition).
Méchain G, Couairon A, Andre Y-B, D'Amico C, Franco M, Prade B, Tzortzakis S, Mysyrowicz A and Sauerbrey R (2004) Appl Phys B 79:379.
Meyerand RG and Haught AF (1963) Phys Rev Lett 11:401.
Meyerand RG and Haught AF (1964) Phys Rev Lett 13:7.

Milosevic DB, Paulus GG, Bauer D and Becker W (2006) J Phys B At Mol Opt Phys 39:R203.
Mlejnek M, Wright EM and Moloney JV (1998) Opt Lett 23:382.
Mlejnek M, Kolesik M, Moloney JV and Wright EM (1999) Phys Rev Lett 83:2938.
Morgan CG (1975) Rep Prog Phys 38:621.
Nibbering ETJ, Curley PF, Grillon G, Prade BS, Franco MA, Salin F and Mysyrowicz A (1996) Opt Lett 21:62.
Pépin H, Comtois D, Vidal F, Chien CY, Desparois A, Johnston TW, Kieffer J-C, La Fontaine B, Martin F, Rizk FAM, Potvin C, Couture P, Mercure HP, Bondiou-Clergerie A, Lalande P and Gallimberti I (2001) Phys Plasma 8:2532.
Reintjes J, Garman RL and Shimizu F (1973) Phys Rev A 8:1486.
Raizer Yu P (1991) Gas Discharge Physics, Springer-Verlag, New York.
Rodriguez M, Sauerbrey R, Wille H, Wõste L, Fuji T, Andr Y-B, Mysyrowicz A, Klingbeil L, Rethmeier K, Kalkner W, Kasparian J, Salmon E, Yu J and Wolf J-P (2002) Opt Lett 27:772.
Rodriguez M, Bourayou R, Méjean G, Kasparian J, Yu J, Salmon E, Scholz A, Stecklum B, Eisloeffel J, Laux U, Hatzes AP., Sauerbrey R, Woeste L and Wolf J-P (2004) Phys Rev E 69:036607.
Rullière C (ed.) (1998) Femtosecond Laser Pulses, Principles and Experiments, Springer, Berlin.
Schroeder H and Chin SL (2004) Opt Commun 234:399–406.
Seideman T (1995) J Chem Phys 103:7887.
Seideman T and Hamilton E (2006) Adv At Mol Opt Phys 52:289.
Silberberg Y (1990) Collapse of optical pulses. Opt Lett 15(22):1282.
Skupin S, Bergé L, Peschel U and Lederer F (2004) Phys Rev Lett 93:023901.
Stapelfeldt H and Seideman T (2003) Rev Mod Phys, 75:543.
Sun Q, Jiang H, Liu Y, Wu Z, Yang H and Gong, H (2005) Opt Lett 30:320–322.
Svanberg S (2004) Atomic and Molecular Spectroscopy: Basic Principles and Practical Applications, 4th ed., Springer, Heidelberg, pp. 287–460.
Talanov VI (1970) Sov Phys JETP Lett 11:199.
Talebpour A, Yang J and Chin SL (1999a) Opt Commun 163:29. Corrections to this reference: There are two printing errors in this paper. In Eq. (2), Z should read Z_{eff}, and in Eq. (8), W_m should be w_m. The × sign should be omitted.
Talebpour A, Petit S and Chin SL (1999b) Opt Commun 171:285.
Talebpour A, Abdel-Fattah M and Chin SL (2000) Opt Commun 183:479.
Talebpour A, Abdel-Fattah M, Bandrauk AD and Chin SL (2001) Laser Phys 11:68.
Tang XP, Becker A, Liu W, Sharifi SM, Kosareva OG, Kandidov VP, Agostini P and Chin SL (2005a) Appl Phys B 80:547–557.
Tang XP, Becker A, Liu W, Sharifi SM, Kosareva OG, Kandidov VP, Agostini P and Chin SL (2005b) Phys Rev A 71:045401.
Théberge F, Liu W, Hosseini SA, Luo Q, Sharifi SM and Chin SL (2005a) Appl Phys B 81:131.
Théberge F, Aközbe N, Liu W, Gravel J-F and Chin SL (2005b) Opt Commun 245:399.
Théberge F, Liu W, Luo Q and Chin SL (2005c) Appl Phys B 80:221.
Théberge F, Liu W, Tr. Simard P, Becke A, Chin SL (2006a) Phys Rev E 74:036406.
Théberge F, Aközbek N, Liu W, Becker A and Chin SL (2006b) Phys Rev Lett 97:023904.
Théberge F (2007) Third-order parametric processes during the filamentation of ultrashort laser pulses in gases. Ph.D. thesis, Université Laval, Québec, Québec, Canada.
Théberge F, Filion J, Akozbek N, Chen Y, Becker A and Chin SL (2007a) Appl Phys B 87:207.
Théberge F, Akõzbek N, Liu W, Filion J and Chin SL (2007b) Opt Commun 276:298.
Tzortzakis S, Prade B, Franco M, Mysyrowicz A, Hüller S and Mora P (2001) Phys Rev E 64:57401.
Varma S, Chen Y-H and Milchberg HM (2008) Phys Rev Lett 101:205001.

Vidal F, Comtois D, Chien CY, Desparois A, La Fontaine B, Johnston TW, Kieffer J-C, Mercure H, Pépin H and Rizk FA (2000) IEEE Trans Plasma Sci 28:418.
Weingartshofer A, Holmes JK, Caudle G, Clarke EM and Krueger H (1977) Phys Rev Lett 39:269.
Weingartshofer A, Clarke EM, Holmes JK and Jung Ch (1979) Phys Rev A 19:2371.
Xie X, Dai J and Zhang X-C (2006) Phys Rev Lett 96:075005.
Xu HL, Daigl J-F, Luo Q, Chin SL (2006a) Appl Phys B 82:655.
Xu HL, Liu W and Chin SL (2006b) Opt Lett 31:1540.
Xu HL, Bernhardt J, Mathieu P, Roy G and Chin SL (2007a) J Appl Phys 101:033124.
Xu HL, Méjean G, Liu W, Kamali Y, Daigle J-F, Azarm A, Simard PT, Mathieu P, Roy G, Simard J-R and Chin SL (2007b) Appl Phys B 87:151.
Xu HL, Kamali Y, Marceau C, Simard PT, Liu W, Bernhardt J, Méjean G, Mathieu P, Roy G, Simard J-R and Chin SL (2007c) Appl Phys Lett 90:101106.
Xu H, Xiong H, Li R, Cheng Y, Xu ZZ and Chin SL (2008) Appl Phys Lett 92:011111.
Yablonovitch E and Bloembergen N (1972) Phys Rev Lett 29:907.

Index

G

Batch number: 10622065

Printed by Printforce, the Netherlands